OFFICIAL SQA PAST PAPERS WITH ANSWERS

STANDARD GRADE | CREDIT

PHYSICS
2006-2010

2006 EXAM – page 3
2007 EXAM – page 31
2008 EXAM – page 59
2009 EXAM – page 93
2010 EXAM – page 127

© Scottish Qualifications Authority
All rights reserved. Copying prohibited. No part of this publication may be reproduced, stored in a retrieval system, or transmitted in any form or by any means, electronic, mechanical, photocopying, recording or otherwise.

First exam published in 2006.
Published by Bright Red Publishing Ltd, 6 Stafford Street, Edinburgh EH3 7AU
tel: 0131 220 5804 fax: 0131 220 6710 info@brightredpublishing.co.uk www.brightredpublishing.co.uk

ISBN 978-1-84948-103-8

A CIP Catalogue record for this book is available from the British Library.

Bright Red Publishing is grateful to the copyright holders, as credited on the final page of the book, for permission to use their material.
Every effort has been made to trace the copyright holders and to obtain their permission for the use of copyright material.
Bright Red Publishing will be happy to receive information allowing us to rectify any error or omission in future editions.

STANDARD GRADE | CREDIT

2006

OFFICIAL SQA PAST PAPERS   5   CREDIT PHYSICS 2006

FOR OFFICIAL USE

C

K & U   PS

Total Marks

3220/402

NATIONAL   WEDNESDAY, 17 MAY   PHYSICS
QUALIFICATIONS   10.50 AM – 12.35 PM   STANDARD GRADE
2006   Credit Level

Fill in these boxes and read what is printed below.

Full name of centre   Town

Forename(s)   Surname

Date of birth
Day Month Year   Scottish candidate number   Number of seat

Reference may be made to the Physics Data Booklet.

1 All questions should be answered.

2 The questions may be answered in any order but all answers must be written clearly and legibly in this book.

3 Write your answer where indicated by the question or in the space provided after the question.

4 If you change your mind about your answer you may score it out and rewrite it in the space provided at the end of the answer book.

5 Before leaving the examination room you must give this book to the invigilator. If you do not, you may lose all the marks for this paper.

6 Any necessary data will be found in the **data sheet** on page two.

7 Care should be taken to give an appropriate number of significant figures in the final answers to questions.

SA 3220/402 6/22770

DATA SHEET

Speed of light in materials

Material	Speed in m/s
Air	3.0×10^8
Carbon dioxide	3.0×10^8
Diamond	1.2×10^8
Glass	2.0×10^8
Glycerol	2.1×10^8
Water	2.3×10^8

Speed of sound in materials

Material	Speed in m/s
Aluminium	5200
Air	340
Bone	4100
Carbon dioxide	270
Glycerol	1900
Muscle	1600
Steel	5200
Tissue	1500
Water	1500

Gravitational field strengths

	Gravitational field strength on the surface in N/kg
Earth	10
Jupiter	26
Mars	4
Mercury	4
Moon	1.6
Neptune	12
Saturn	11
Sun	270
Venus	9

Specific heat capacity of materials

Material	Specific heat capacity in J/kg °C
Alcohol	2350
Aluminium	902
Copper	386
Diamond	530
Glass	500
Glycerol	2400
Ice	2100
Lead	128
Water	4180

Specific latent heat of fusion of materials

Material	Specific latent heat of fusion in J/kg
Alcohol	0.99×10^5
Aluminium	3.95×10^5
Carbon dioxide	1.80×10^5
Copper	2.05×10^5
Glycerol	1.81×10^5
Lead	0.25×10^5
Water	3.34×10^5

Melting and boiling points of materials

Material	Melting point in °C	Boiling point in °C
Alcohol	−98	65
Aluminium	660	2470
Copper	1077	2567
Glycerol	18	290
Lead	328	1737
Turpentine	−10	156

Specific latent heat of vaporisation of materials

Material	Specific latent heat of vaporisation in J/kg
Alcohol	11.2×10^5
Carbon dioxide	3.77×10^5
Glycerol	8.30×10^5
Turpentine	2.90×10^5
Water	22.6×10^5

SI Prefixes and Multiplication Factors

Prefix	Symbol	Factor	
giga	G	1 000 000 000	$= 10^9$
mega	M	1 000 000	$= 10^6$
kilo	k	1000	$= 10^3$
milli	m	0.001	$= 10^{-3}$
micro	µ	0.000 001	$= 10^{-6}$
nano	n	0.000 000 001	$= 10^{-9}$

1. A computer is connected to the Internet by means of a copper wire and a glass optical fibre as shown.

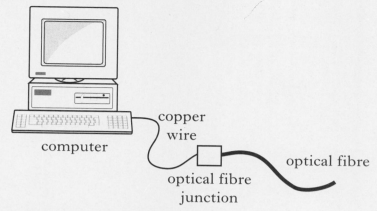

(a) In the table below, enter:

  (i) the speed of the signal in each material;

  (ii) the type of signal in each material.

	Copper wire	Glass optical fibre
Speed of signal		
Type of signal		

(b) Complete the diagram to show how the signal travels along the optical fibre.

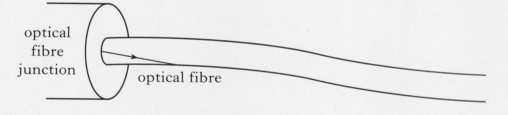

(c) Copper wire or glass optical fibre can be used in telecommunication systems.

  (i) Explain which material, copper or glass, would need less repeater amplifiers over a long distance.

  ....

  ....

  (ii) A broadband communication system carries 100 television channels and 200 phone channels.

  Explain which material, copper or glass, should be used in this system.

  ....

  ....

2. A ship has a satellite navigation system. A receiver on the ship picks up signals from three global positioning satellites.

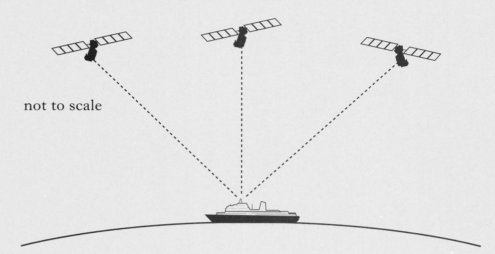

not to scale

These satellites can transmit radio signals at three different frequencies, 1176 MHz, 1228 MHz and 1575 MHz. The satellites orbit at a height of 20 200 km above the Earth's surface.

(a) (i) State the speed of the radio signals.

....

(ii) One of the satellites is directly above the ship.

Calculate the time taken for the signal from this satellite to reach the ship.

Space for working and answer

(iii) Calculate the wavelength of the 1228 MHz signal.

Space for working and answer

2. (continued)

(b) State which of the three signals has the shortest wavelength.

.... 1

(c) One of the global positioning satellites is shown below.

(i) Complete the diagram below to show the effect of the curved reflector on the transmitted signals.

2

(ii) A satellite in orbit a few hundred kilometres above Earth has a period of one hour. A geostationary satellite orbits 36 000 km above Earth.

Suggest the period of the global positioning satellite.

.... 1

[Turn over

3. Two students are investigating voltage, current and resistance.

(a) The first student builds the circuit shown.

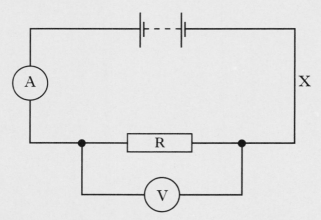

The ammeter displays a current of 0·10 A and the voltmeter displays a voltage of 3·0 V.

(i) Calculate the resistance of R when the current is 0·10 A.

Space for working and answer

(ii) The student inserts another ammeter at position X.

What is the reading on this ammeter?

....

(b) The second student uses the **same** resistor in the circuit below.

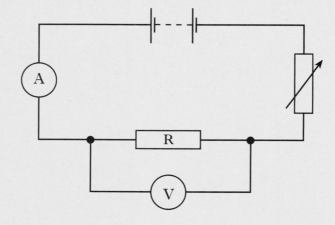

3. *(b)* **(continued)**

This student obtains the following set of results.

Result number	Voltage across R (V)	Current through R (A)
1	6·0	0·20
2	7·5	0·25
3	9·0	0·30
4	10·0	0·35
5	12·0	0·40

(i) Describe how these different values of voltage and current are obtained.

....

.... **2**

(ii) Explain which result should be retaken.

....

.... **2**

(c) What additional information about resistance does the second student's experiment give compared to the first student's experiment?

....

....

.... **1**

[*Turn over*

4. A circuit breaker as shown below is used in a circuit.

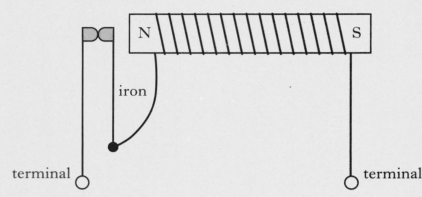

(a) (i) State **one** advantage of a circuit breaker compared to a fuse.

....

.... 1

(ii) The circuit breaker breaks the circuit when the current becomes too high.

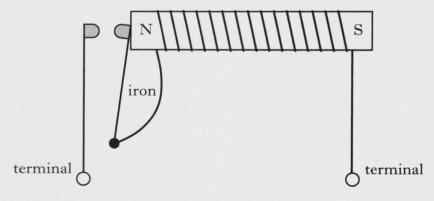

Explain how the circuit breaker operates when the current becomes too high.

....

....

.... 2

4. (continued)

(b) A 5 ampere circuit breaker is used in a household lighting circuit which has three 60 W lamps as shown below.

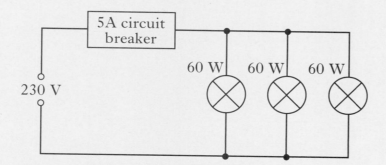

(i) Show that the resistance of **one** lamp is 882 Ω.

Space for working and answer

(ii) Calculate the combined resistance of the three lamps in this circuit.

Space for working and answer

(iii) Show by calculation whether the circuit breaker will switch off the lamps when all three are lit.

Space for working and answer

5. A radioactive source is used for medical treatment. The graph shows the activity of this source over a period of time.

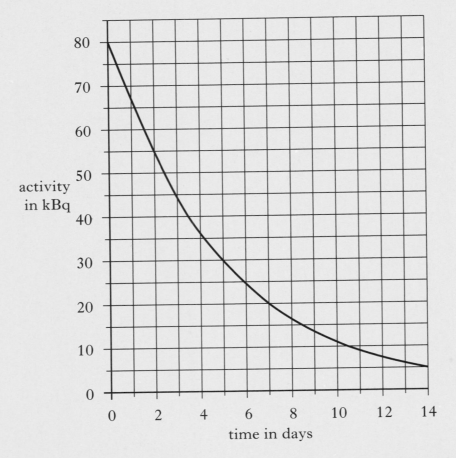

(a) Use information from the graph to calculate the half-life of this source.

Space for working and answer

5. (continued)

(b) Describe a method that could be used to measure the half-life of this radioactive source, using the apparatus shown. You can ignore background radiation.

....
....
....
....
....

2

(c) A sample of this source is to be given to a patient at 9.30 am on May 17. When the sample is prepared, its initial activity is 200 kBq. The activity of the sample when given to the patient must be 12·5 kBq.

Calculate at what time and on what date the sample should be prepared.

Space for working and answer

2

[Turn over

6. The table below gives information about some types of laser.

Type of laser	Wavelength (nm)	Output power (W)
Krypton fluoride	248	1·0
Argon	488	2·0
Helium neon	633	0·005
Rhodamine	570 to 650	50·0
Carbon dioxide	10 600	200·0

(a) The visible spectrum has wavelengths ranging from 400 nm to 700 nm.

  (i) Name one type of laser **from the table** that emits visible radiation.

  ....

  (ii) Name one type of laser **from the table** that emits ultraviolet radiation.

  ....

  (iii) Give **one** medical use of ultraviolet radiation.

  ....

(b) A rhodamine laser can be adjusted to emit a range of wavelengths.

What difference is observed in the light from this laser beam as the wavelength changes?

....

....

(c) The beam from the carbon dioxide laser is used to cut steel. A section of steel is cut in 10 minutes.

Using information from the table, calculate the energy given out by the laser during this cutting process.

Space for working and answer

7. A student designs a lie detector using the following circuit.

Moisture detector:
high resistance when dry
low resistance when wet

(a) Name component Q.

.... 1

(b) Suggest a suitable output device that could be used at P to produce an audible output.

.... 1

(c) This lie detector is based on the fact that when a person tells a lie, the moisture on their skin increases. Initially, the person holds the moisture detector in dry hands and component R is adjusted until the output device is silent.

  (i) What happens to the resistance of the moisture detector when the person holding it tells a lie?

  ....

  1

  (ii) Explain how the circuit operates as a lie detector.

  ....

  ....

  ....

  2

[Turn over

8. An automatic vending machine accepts 1p, 2p and 5p coins. Four light sensors P, Q, R and S are arranged as shown in the coin slot.

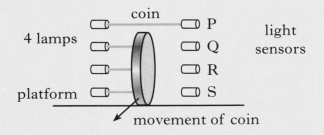

When a coin passes between a lamp and its sensor, the light is blocked. Coins of different diameters block the light from different lamps.

The position of the sensors in relation to the diameters of coins is shown below.

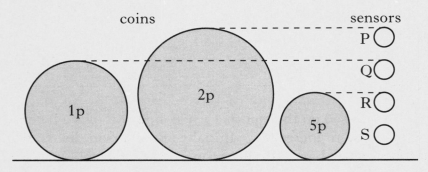

The logic output of the sensors is as follows:

light blocked — logic output 1
light not blocked — logic output 0

(a) (i) Name a suitable input device to be used as a sensor.

....

(ii) Complete the truth table for the outputs of the sensors when each of the coins passes between the lamps and the sensors.

	1p coin	2p coin	5p coin
Sensor P			
Sensor Q			
Sensor R			
Sensor S			

8. (continued)

(b) A washer is a metal disc with a hole in the middle. The machine is able to reject washers, when they are inserted instead of coins. A washer the same diameter as a 1p coin blocks the light from reaching sensors Q and S only.

Part of the circuit used is shown below.

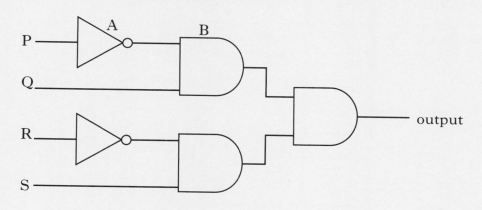

(i) Name gate A.

.... 1

(ii) Name gate B.

.... 1

(iii) When a washer is inserted, the logic levels at P, Q, R and S are as shown below.

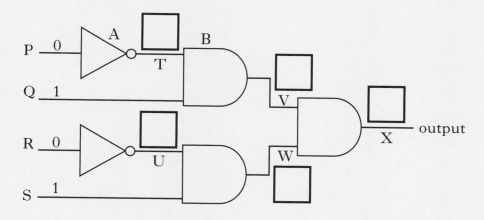

In the boxes on the diagram above, enter the logic levels at each position T, U, V, W and X. 2

(iv) When a washer is detected, this circuit activates an output device that pushes the washer to reject it.

Name a suitable device to be used as the output device.

.... 1

9. A table from the Highway Code giving overall stopping distances for vehicles is shown.

The overall stopping distance is made up of:

the **thinking distance** – the distance travelled while the driver "thinks" about braking. This distance depends on the driver's reaction time.
plus
the **braking distance** – the distance travelled while braking.

Speed of vehicle (m/s)	Overall stopping distance (m)
8·9	6 + 6
13·4	9 + 14
17·8	12 + 24
26·7	18 + 55

thinking distance | braking distance

(a) (i) How far does a vehicle travelling at 13·4 m/s travel while the driver thinks about braking?

.... **1**

(ii) Use information **from the table** to calculate the reaction time.

Space for working and answer

2

9. (continued)

(b) A car travels along a road. The driver sees traffic lights ahead change from green and starts to brake as soon as possible. A graph of the car's motion, from the moment the driver sees the traffic lights change, is shown.

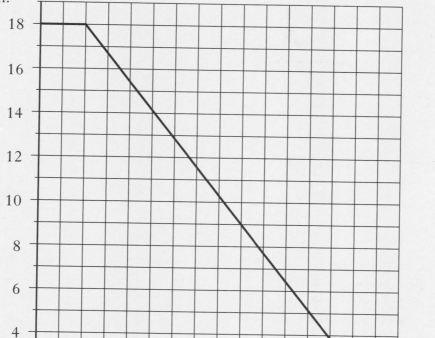

(i) What is **this** driver's reaction time?

.... 1

(ii) Calculate the overall stopping distance.

Space for working and answer

3

(iii) Calculate the acceleration of the car from the time the driver applies the brakes.

Space for working and answer

2

10. A student runs along a diving platform and leaves the platform horizontally with a speed of 2·0 m/s. The student lands in the water 0·3 s later. Air resistance is negligible.

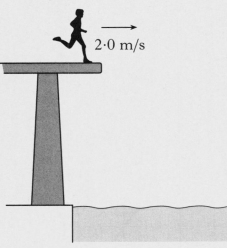

(a) (i) Calculate the horizontal distance travelled by the student before landing in the water.

Space for working and answer

(ii) The student has a vertical acceleration of 10 m/s^2.

Calculate the vertical speed as the student enters the water.

Space for working and answer

(b) Later the student runs off the end of the same platform with a horizontal speed of 3·0 m/s.

How long does the student take to reach the water this time? Explain your answer.

10. (continued)

(c) The student climbs from the water level to a higher platform. This platform is 5·0 m above the water. The student has a mass of 50 kg.

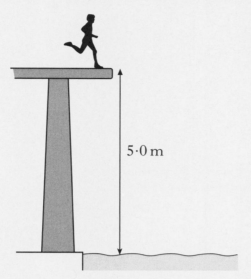

(i) Calculate the gain in gravitational potential energy of the student.

Space for working and answer

(ii) The student drops from the edge of the platform and lands in the water.

Calculate the vertical speed as the student enters the water.

Space for working and answer

[Turn over

11. A wind generator on a yacht is used to charge a battery at 12 V.

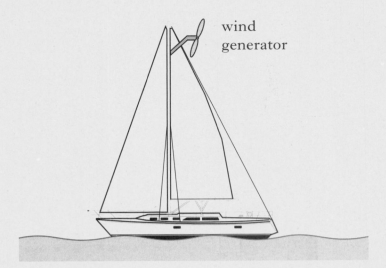

The graph shows the charging current at different wind speeds.

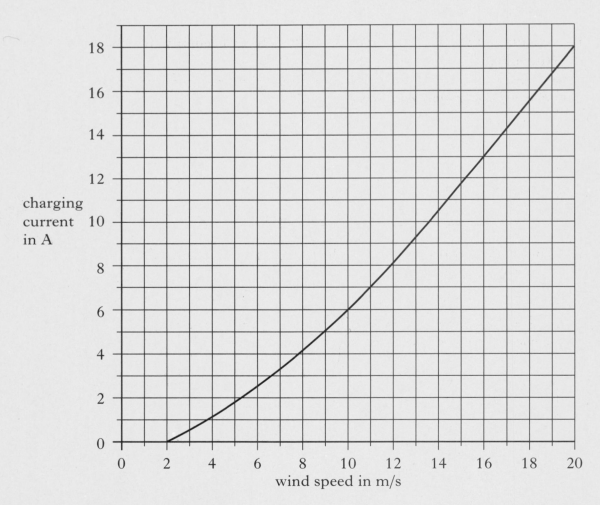

(a) The wind blows at a speed of 10 m/s.
  (i) What is the charging current at this wind speed?

....

11. (*a*) (**continued**)

(ii) Calculate the electrical power produced by the generator at this wind speed.

Space for working and answer

(iii) The wind speed does not change.

Calculate the energy supplied to the battery in 3·5 hours.

Space for working and answer

(*b*) The yacht has a stand-by petrol powered generator to charge the battery.

Why is the petrol generator necessary, in addition to the wind generator?

....

....

[Turn over

12. A mains operated air heater contains a fan, driven by a motor, and a heating element. Cold air is drawn into the heater by the fan. The air is heated as it passes the heating element.

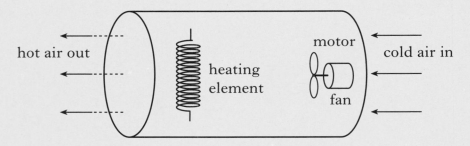

The circuit diagram for the air heater is shown.

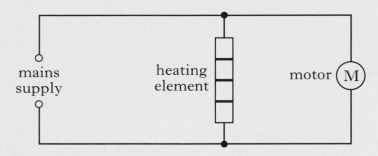

(a) (i) What is the voltage across the heating element when the heater is operating?

....

(ii) What type of circuit is used for the air heater?

....

(b) The following data relates to the heater when the fan rotates at a particular speed.

    mass of air passing through per second   0·2 kg
    energy supplied to air per second   2000 J
    specific heat capacity of air   1000 J/kg °C

(i) Calculate the increase in air temperature.

Space for working and answer

12. (b) (continued)

(ii) The motor is adjusted to rotate the fan at a higher speed. This draws a greater mass of air per second through the heater. Explain any difference this causes to the temperature of the hot air.

....

....

.... **2**

[Turn over

13. Titan is the largest of Saturn's moons. The gravitational field strength on Titan is 1·35 N/kg.

(a) (i) What is a moon?

....

....

(ii) What is meant by gravitational field strength?

....

....

(b) Early in 2005, a probe was released from a spacecraft orbiting Titan. The probe, of mass 318 kg, travelled through the atmosphere of Titan.

(i) Calculate the weight of the probe on Titan.

Space for working and answer

(ii) As the probe descended through the atmosphere, a parachute attached to it opened.

State why the parachute was used.

....

....

....

13. (b) (continued)

(iii) The probe carried equipment to analyse the spectral lines of radiation from gases in the atmosphere of Titan. These lines are shown. The spectral lines of a number of elements are also shown.

Spectral lines from gases in Titan's atmosphere

Helium

Hydrogen

Mercury

Nitrogen

Use the spectral lines of the elements to identify which elements are present in the atmosphere of Titan.

....

....

....

[END OF QUESTION PAPER]

YOU MAY USE THE SPACE ON THIS PAGE TO REWRITE ANY ANSWER YOU HAVE DECIDED TO CHANGE IN THE MAIN PART OF THE ANSWER BOOKLET. TAKE CARE TO WRITE IN CAREFULLY THE APPROPRIATE QUESTION NUMBER.

STANDARD GRADE | CREDIT
2007

[BLANK PAGE]

FOR OFFICIAL USE

K&U | PS

Total Marks

3220/402

NATIONAL QUALIFICATIONS 2007

WEDNESDAY, 16 MAY 10.50 AM – 12.35 PM

PHYSICS STANDARD GRADE Credit Level

Fill in these boxes and read what is printed below.

Full name of centre

Town

Forename(s)

Surname

Date of birth
Day Month Year

Scottish candidate number

Number of seat

Reference may be made to the Physics Data Booklet.

1 All questions should be answered.

2 The questions may be answered in any order but all answers must be written clearly and legibly in this book.

3 Write your answer where indicated by the question or in the space provided after the question.

4 If you change your mind about your answer you may score it out and rewrite it in the space provided at the end of the answer book.

5 Before leaving the examination room you must give this book to the invigilator. If you do not, you may lose all the marks for this paper.

6 Any necessary data will be found in the **data sheet** on page two.

7 Care should be taken to give an appropriate number of significant figures in the final answers to questions.

SA 3220/402 6/22870

DATA SHEET

Speed of light in materials

Material	Speed in m/s
Air	3.0×10^8
Carbon dioxide	3.0×10^8
Diamond	1.2×10^8
Glass	2.0×10^8
Glycerol	2.1×10^8
Water	2.3×10^8

Speed of sound in materials

Material	Speed in m/s
Aluminium	5200
Air	340
Bone	4100
Carbon dioxide	270
Glycerol	1900
Muscle	1600
Steel	5200
Tissue	1500
Water	1500

Gravitational field strengths

	Gravitational field strength on the surface in N/kg
Earth	10
Jupiter	26
Mars	4
Mercury	4
Moon	1.6
Neptune	12
Saturn	11
Sun	270
Venus	9

Specific heat capacity of materials

Material	Specific heat capacity in J/kg °C
Alcohol	2350
Aluminium	902
Copper	386
Diamond	530
Glass	500
Glycerol	2400
Ice	2100
Lead	128
Water	4180

Specific latent heat of fusion of materials

Material	Specific latent heat of fusion in J/kg
Alcohol	0.99×10^5
Aluminium	3.95×10^5
Carbon dioxide	1.80×10^5
Copper	2.05×10^5
Glycerol	1.81×10^5
Lead	0.25×10^5
Water	3.34×10^5

Melting and boiling points of materials

Material	Melting point in °C	Boiling point in °C
Alcohol	−98	65
Aluminium	660	2470
Copper	1077	2567
Glycerol	18	290
Lead	328	1737
Turpentine	−10	156

Specific latent heat of vaporisation of materials

Material	Specific latent heat of vaporisation in J/kg
Alcohol	11.2×10^5
Carbon dioxide	3.77×10^5
Glycerol	8.30×10^5
Turpentine	2.90×10^5
Water	22.6×10^5

SI Prefixes and Multiplication Factors

Prefix	Symbol	Factor	
giga	G	1 000 000 000	$= 10^9$
mega	M	1 000 000	$= 10^6$
kilo	k	1000	$= 10^3$
milli	m	0.001	$= 10^{-3}$
micro	μ	0.000 001	$= 10^{-6}$
nano	n	0.000 000 001	$= 10^{-9}$

1. A pupil is sent exam results by a text message on a mobile phone. The frequency of the signal received by the phone is 1900 MHz.

The mobile phone receives radio waves (signals).

(a) What is the speed of radio waves?

....

(b) Calculate the wavelength of the signal.

Space for working and answer

(c) The pupil sends a video message from the mobile phone. The message is transmitted by microwaves. The message travels a total distance of 72 000 km.

Calculate the time between the message being transmitted and received.

Space for working and answer

2. Radio waves have a wide range of frequencies.

The table gives information about different wavebands.

Waveband	Frequency Range	Example
Low frequency (LF)	30 kHz – 300 kHz	Radio 4
Medium frequency (MF)	300 kHz – 3 MHz	Radio Scotland
High frequency (HF)	3 MHz – 30 MHz	Amateur radio
Very high frequency (VHF)	30 MHz – 300 MHz	Radio 1 FM
Ultra high frequency (UHF)	300 MHz – 3 GHz	BBC 1 and ITV
Super high frequency (SHF)	3 GHz – 30 GHz	Satellite TV

(a) Coastguards use signals of frequency 500 kHz.

What waveband do these signals belong to?

.... 1

2. (continued)

(b) The diagram shows how radio signals of different wavelengths are sent between a transmitter and a receiver.

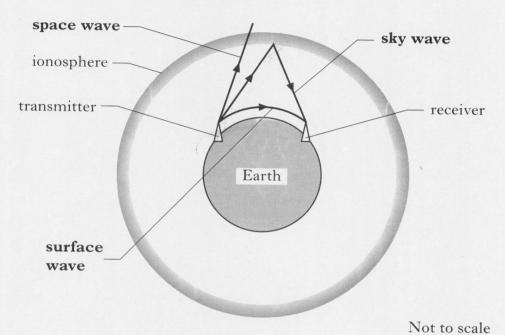

Not to scale

(i) Which of the waves in the diagram shows diffraction?

.... **1**

(ii) What does this indicate about the wavelength of the diffracted wave compared to the other two waves?

.... **1**

(iii) The Earth's ionosphere is shown on the diagram. The ionosphere is a layer of charged particles in the upper atmosphere. High frequency waves are transmitted as sky waves.

Explain how the transmitted waves reach the receiver.

.... **1**

(iv) Super high frequency (SHF) signals are shown as space waves on the diagram. Although they can only travel in straight lines, they can be used for communications on Earth between a transmitter and receiver.

Describe how the SHF signals get to the receiver.

....

....

.... **2**

3. A door entry system in an office block allows video and audio information to be sent between two people.

(a) A camera at the entrance uses a lens to focus parallel rays of light onto a detector.

Part of the camera is shown in the diagram below.

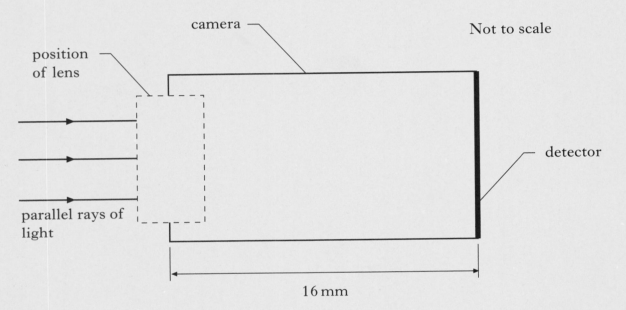

(i) Complete the diagram above by:

(A) drawing the lens used;

(B) completing the path of the light rays.

(ii) Using information from the diagram, calculate the power of the lens used in the camera.

Space for working and answer

3. (continued)

(b) The door entry system uses a black and white television screen.

Describe how a moving picture is seen on the television screen.

Your description must include the terms:

line build up   image retention   brightness variation.

....

....

....

....

....

3

[Turn over

4. The consumer unit in a house contains a mains switch and circuit breakers for different circuits.

(a) (i) What is the purpose of the mains switch?

.... 1

(ii) Two of the circuits have not been labelled.

Which circuit is: the ring circuit?   ....

the lighting circuit?   1

(iii) The current ratings for the ring circuit and the lighting circuit are different.

State another difference between the ring circuit and the lighting circuit.

.... 1

DO NOT WRITE IN THIS MARGIN

Marks | K&U | PS

4. (continued)

(*b*) (i) A 25 W lamp is designed to be used with mains voltage.

Calculate the resistance of the lamp.

Space for working and answer

3

(ii) Four of these lamps are connected in parallel.

Calculate the **total** resistance of the lamps.

Space for working and answer

2

[Turn over

5. Two groups of pupils are investigating the electrical properties of a lamp.

(a) Group 1 is given the following equipment:

ammeter; voltmeter; 12 V d.c. supply; lamp; connecting leads.

Complete the circuit diagram to show how this equipment is used to measure the current through, and the voltage across, the lamp.

———o 12 V d.c. o———

(b) Group 2 uses the same lamp and is only given the following equipment:

lamp; ohmmeter; connecting leads.

What property of the lamp is measured by the ohmmeter?

....

(c) The results of both groups are combined and recorded in the table below.

I(A)	V(V)	R(Ω)	IV	I^2R
2	12	6		

(i) Use these results to complete the last two columns of the table.

Space for working

(ii) What quantity is represented by the last two columns of the table?

....

(iii) What is the unit for this quantity?

....

6. The thyroid gland, located in the neck, is essential for maintaining good health.

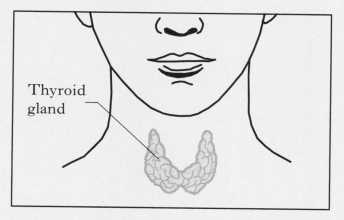

(a) (i) A radioactive source, which is a gamma radiation emitter, is used as a radioactive tracer for the diagnosis of thyroid gland disorders.

A small quantity of this tracer, with an activity of 20 MBq, is injected into a patient's body. After 52 hours, the activity of the tracer is measured at 1·25 MBq.

Calculate the half life of the tracer.

Space for working and answer

(ii) Another radioactive source is used to **treat** cancer of the thyroid gland. This source emits only beta radiation.

Why is this source unsuitable as a **tracer**?

....

....

(iii) The equivalent dose is much higher for the beta emitter than for the gamma emitter.

Why is this higher dose necessary?

....

(b) What are the units of equivalent dose?

....

7. A newborn baby is given a hearing test. A small device, containing a loudspeaker and a microphone, is placed in the baby's ear.

(a) A pulse of audible sound lasting 10 μs is transmitted through the loudspeaker. The sound is played at a level of 80 dB.

(i) Give a reason why this pulse of sound does not cause damage to the baby's hearing.

....

.... 1

7. (a) (continued)

(ii) The transmitted pulse of sound makes the inner ear vibrate to produce a new sound, which is received by the microphone.

Signals from the transmitted and received sounds are viewed on an oscilloscope screen, as shown below.

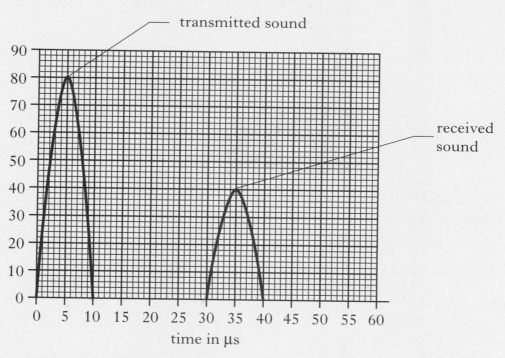

The average speed of sound inside the ear is 1500 m/s.

Calculate the distance between the device and the inner ear.

Space for working and answer

(iii) Suggest a frequency that could be used for the hearing test.

....

(b) An ultrasound scan can be used to produce an image of an unborn baby. Explain how the image of an unborn baby is formed by ultrasound.

....

....

....

8. A high intensity LED is used as a garden light. The light turns on automatically when it becomes dark.

The light also contains a solar cell which charges a rechargeable battery during daylight hours.

(a) Part of the circuit is shown below.

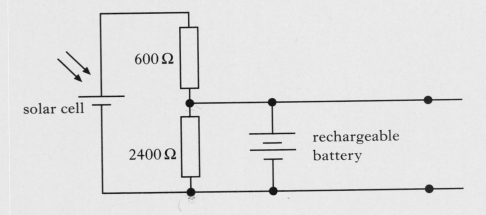

(i) State the energy transformation in a solar cell.

....

(ii) At a particular light level, the voltage generated by the solar cell is 1·5 V.

Calculate the voltage across the rechargeable battery at this light level.

Space for working and answer

8. (continued)

(b) The LED is switched on using the following circuit.

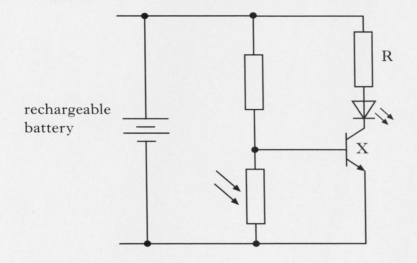

(i) Name component X.

.... 1

The graph below shows the voltage across the LDR in this circuit for different light levels.

Light level is measured in lux.

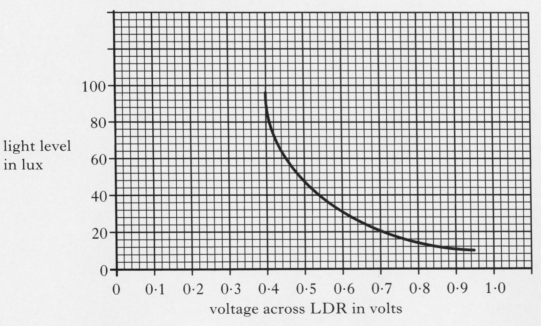

(ii) For the LED to be lit, the voltage across the LDR must be at least 0·7 V.

What is the maximum light level for the LED to be lit?

.... 1

(iii) Explain the purpose of resistor R.

.... 1

9. An electronic tuner for a guitar contains a microphone and an amplifier. The output voltage from the amplifier is 9 V.

(a) The voltage gain of the amplifier is 150.

Calculate the input voltage to the amplifier.

Space for working and answer

(b) The tuner is used to measure the frequency of six guitar strings.

The number and frequency of each string is given in the table below.

Number of string	Frequency (Hz)
1	330·0
2	247·0
3	196·0
4	147·0
5	110·0
6	82·5

The tuner has an output socket which has been connected to an oscilloscope. The trace for string 5 is shown in Figure 1.

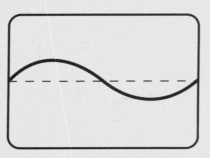

Figure 1

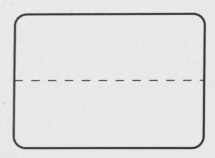

Figure 2

(i) The controls of the oscilloscope are **not** altered.

In Figure 2, draw the trace obtained if string 1 is played **louder** than string 5.

(ii) String 3 is plucked.

What is the frequency of the output signal from the amplifier?

10. Cameras placed at 5 km intervals along a stretch of road are used to record the average speed of a car.

The car is travelling on a road which has a speed limit of 100 km/h. The car travels a distance of 5 km in 2·5 minutes.

(a) Does the average speed of the car stay within the speed limit?

You must justify your answer with a calculation.

Space for working and answer

(b) At one point in the journey, the car speedometer records 90 km/h.

Explain why the average speed for the entire journey is not always the same as the speed recorded on the car speedometer.

[Turn over

11. An aeroplane on an aircraft carrier must reach a minimum speed of 70 m/s to safely take off. The mass of the aeroplane is 28 000 kg.

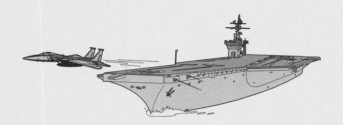

(a) The aeroplane accelerates from rest to its minimum take off speed in 2 s.

  (i) Calculate the acceleration of the aeroplane.

  Space for working and answer

  (ii) Calculate the force required to produce this acceleration.

  Space for working and answer

  (iii) The aeroplane's engines provide a total thrust of 240 kN. An additional force is supplied by a catapult to produce the acceleration required.

  Calculate the force supplied by the catapult.

  Space for working and answer

11. (continued)

(b) Later, the same aeroplane travelling at a speed of 65 m/s, touches down on the carrier.

(i) Calculate the kinetic energy of the aeroplane at this speed.

Space for working and answer

(ii) The graph shows the motion of the aeroplane from the point when it touches down on the carrier until it stops.

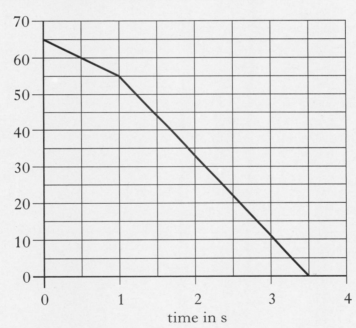

Calculate the distance travelled by the aeroplane on the carrier.

Space for working and answer

12. The advertisement below is for a new torch.

> **Kinetic Torch**
> No batteries needed – magnet powered!
> Bright white LED won't burn out!
> 30-40 seconds of gentle shaking produces 10-15 minutes of light!
> Capacitor holds the charge generated by passing the magnet through the coil.

(a) (i) Explain how a voltage is induced in the coil.

....

.... 2

(ii) What is the effect of shaking the torch faster?

.... 1

(iii) Draw the circuit symbol for a capacitor.

Space for symbol

1

(b) When lit, the current in the LED is 20 mA.

Calculate how much charge flows through the LED in 12 minutes.

Space for working and answer

2

12. (continued)

(c) The torch produces a beam of light.

The diagram shows the LED positioned at the focus of the torch reflector.

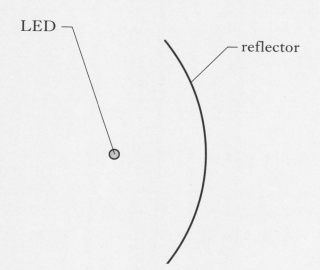

Complete the diagram by drawing light rays to show how the beam of light is produced.

2

[Turn over

13. An electric kettle is used to heat 0·4 kg of water.

(a) The initial temperature of the water is 15 °C.

Calculate how much heat energy is required to bring this water to its boiling point of 100 °C.

Space for working and answer

(b) The automatic switch on the kettle is not working. The kettle is switched off 5 minutes after it had been switched on.

The power rating of the kettle is 2000 W.

(i) Calculate how much electrical energy is converted into heat energy in this time.

Space for working and answer

(ii) Calculate the mass of water changed into steam in this time.

Space for working and answer

14. The diagram represents the electromagnetic spectrum in order of increasing wavelength. Some of the radiations have not been named.

Electromagnetic Spectrum

| Gamma rays | P | Ultraviolet | Q | Infrared | R | TV and Radio |

→ increasing wavelength

(a) (i) Name radiation: P ....

Q ....

R ....

(ii) Which radiation in the electromagnetic spectrum has the highest frequency?

....

(b) Stars emit **ultraviolet** and **infrared** radiation.

Name a detector for **each** of these two radiations.

Infrared ....

Ultraviolet ....

[Turn over

15. In June 2005, a space vehicle called Mars Lander was sent to the planet Mars.

(a) The graph shows the gravitational field strength at different heights above the surface of Mars.

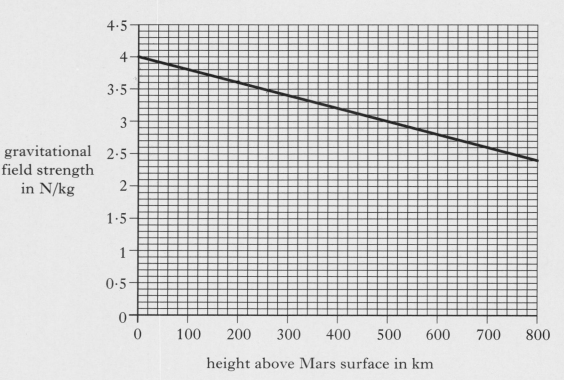

(i) The Mars Lander orbited Mars at a height of 200 km above the planet's surface.

What is the value of the gravitational field strength at this height?

....

(ii) The Mars Lander, of mass 530 kg, then landed.

Calculate the weight of the Mars Lander on the surface.

Space for working and answer

15. (continued)

(b) The Mars Lander released a rover exploration vehicle on to the surface of Mars.

To collect data from the bottom of a large crater, the rover launched a probe horizontally at 30 m/s. The probe took 6 s to reach the bottom of the crater.

(i) Calculate the horizontal distance travelled by the probe.

Space for working and answer

(ii) Calculate the vertical speed of the probe as it reached the bottom of the crater.

Space for working and answer

[END OF QUESTION PAPER]

YOU MAY USE THE SPACE ON THIS PAGE TO REWRITE ANY ANSWER YOU HAVE DECIDED TO CHANGE IN THE MAIN PART OF THE ANSWER BOOKLET. TAKE CARE TO WRITE IN CAREFULLY THE APPROPRIATE QUESTION NUMBER.

STANDARD GRADE | CREDIT

2008

OFFICIAL SQA PAST PAPERS   61   CREDIT PHYSICS 2008

FOR OFFICIAL USE

C

K & U | PS

Total Marks

3220/402

NATIONAL QUALIFICATIONS 2008

FRIDAY, 23 MAY 10.50 AM – 12.35 PM

PHYSICS
STANDARD GRADE
Credit Level

Fill in these boxes and read what is printed below.

Full name of centre

Town

Forename(s)

Surname

Date of birth
Day Month Year   Scottish candidate number   Number of seat

Reference may be made to the Physics Data Booklet.

1 All questions should be answered.

2 The questions may be answered in any order but all answers must be written clearly and legibly in this book.

3 Write your answer where indicated by the question or in the space provided after the question.

4 If you change your mind about your answer you may score it out and rewrite it in the space provided at the end of the answer book.

5 Before leaving the examination room you must give this book to the invigilator. If you do not, you may lose all the marks for this paper.

6 Any necessary data will be found in the **data sheet** on page three.

7 Care should be taken to give an appropriate number of significant figures in the final answers to questions.

SA 3220/402 6/21770

[BLANK PAGE]

DATA SHEET

Speed of light in materials

Material	Speed in m/s
Air	3.0×10^8
Carbon dioxide	3.0×10^8
Diamond	1.2×10^8
Glass	2.0×10^8
Glycerol	2.1×10^8
Water	2.3×10^8

Speed of sound in materials

Material	Speed in m/s
Aluminium	5200
Air	340
Bone	4100
Carbon dioxide	270
Glycerol	1900
Muscle	1600
Steel	5200
Tissue	1500
Water	1500

Gravitational field strengths

	Gravitational field strength on the surface in N/kg
Earth	10
Jupiter	26
Mars	4
Mercury	4
Moon	1.6
Neptune	12
Saturn	11
Sun	270
Venus	9

Specific heat capacity of materials

Material	Specific heat capacity in J/kg °C
Alcohol	2350
Aluminium	902
Copper	386
Glass	500
Glycerol	2400
Ice	2100
Lead	128
Silica	1033
Water	4180

Specific latent heat of fusion of materials

Material	Specific latent heat of fusion in J/kg
Alcohol	0.99×10^5
Aluminium	3.95×10^5
Carbon dioxide	1.80×10^5
Copper	2.05×10^5
Glycerol	1.81×10^5
Lead	0.25×10^5
Water	3.34×10^5

Melting and boiling points of materials

Material	Melting point in °C	Boiling point in °C
Alcohol	−98	65
Aluminium	660	2470
Copper	1077	2567
Glycerol	18	290
Lead	328	1737
Turpentine	−10	156

Specific latent heat of vaporisation of materials

Material	Specific latent heat of vaporisation in J/kg
Alcohol	11.2×10^5
Carbon dioxide	3.77×10^5
Glycerol	8.30×10^5
Turpentine	2.90×10^5
Water	22.6×10^5

SI Prefixes and Multiplication Factors

Prefix	Symbol	Factor	
giga	G	1 000 000 000	$= 10^9$
mega	M	1 000 000	$= 10^6$
kilo	k	1000	$= 10^3$
milli	m	0.001	$= 10^{-3}$
micro	μ	0.000 001	$= 10^{-6}$
nano	n	0.000 000 001	$= 10^{-9}$

1. A high definition television picture has 1080 lines and there are 25 pictures produced each second.

(a) (i) Calculate how long it takes to produce one picture on the screen.

Space for working and answer

(ii) Explain why a continuous moving picture is seen on the television screen and not 25 individual pictures each second.

....

....

....

(b) The television picture is in colour.

(i) Which **two** colours are used to produce magenta on the screen?

....

(ii) Due to a fault, the colour yellow appears as orange on the screen. Which colour should be reduced in brightness to correct this problem?

....

2. A television company is making a programme in China.

Britain receives television pictures live from China. The television signals are transmitted using microwaves. The microwave signals travel from China **via** a satellite, which is in a geostationary orbit.

(a) State what is meant by a geostationary orbit.

.... 1

(b) The diagram shows the position of the transmitter and receiver. Complete the diagram to show the path of the microwave signals **from** China **to** Britain.

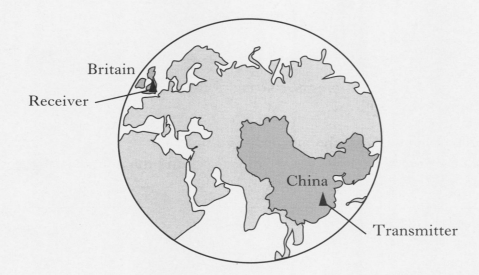

2

(c) The frequency of the microwave signals being used for transmission is 8 GHz.

(i) What is the speed of the microwaves?

.... 1

(ii) Calculate the wavelength of these microwaves.

Space for working and answer

2

3. In a sprint race at a school sports day, the runners start when they hear the sound of the starting pistol. An electronic timer is also started when the pistol is fired into the air.

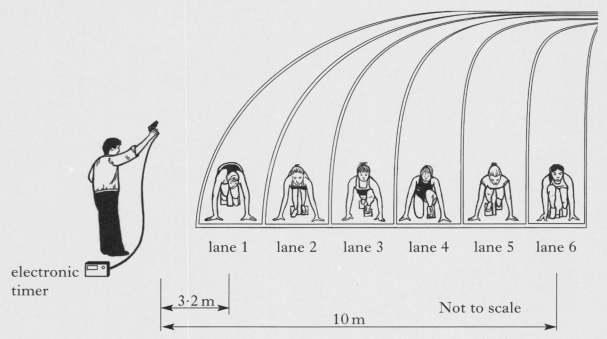

The runner in lane 1 is 3·2 m from the starting pistol. The runner in lane 6 is 10 m from the starting pistol.

(a) The runner in lane 1 hears the starting pistol first.

Calculate how much later the runner in lane 6 hears this sound after the runner in lane 1.

Space for working and answer

3

3. **(continued)**

(b) A sensor detects each runner crossing the finishing line to record their time.

The table gives information about the race.

Place	Lane	Time (s)
1st	1	13·11
2nd	6	13·12
3rd	3	13·21

Using your answer to part (a), explain why the runner in lane 6 should have been awarded first place.

Space for working and answer

2

(c) One runner of mass 60 kg has a speed of 9 m/s when crossing the finishing line.

Calculate the kinetic energy of the runner at this point.

Space for working and answer

2

[Turn over

4. A student has four resistors labelled A, B, C and D. The student sets up Circuit 1 to identify the value of each resistor.

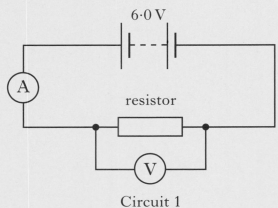

Circuit 1

Each resistor is placed in the circuit in turn and the following results are obtained.

Resistor	Voltage across resistor (V)	Current (A)
A	6·0	0·017
B	6·0	0·027
C	6·0	0·050
D	6·0	0·033

(a) (i) Show, **by calculation**, which of the resistors has a value of 120 Ω.

Space for working and answer

3

4. (a) (continued)

(ii) The student then sets up Circuit 2 to measure the resistance of each resistor.

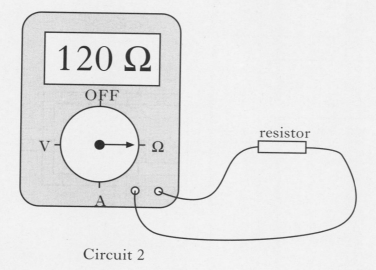

Circuit 2

State **one** advantage of using Circuit 2 to measure the resistance compared to using Circuit 1.

.... 1

(b) The resistances of the other three resistors are 180 Ω, 220 Ω and 360 Ω. The student connects all four resistors in series.

Calculate the total resistance.

Space for working and answer

2

[Turn over

5. The diagram shows three household circuits connected to a consumer unit.

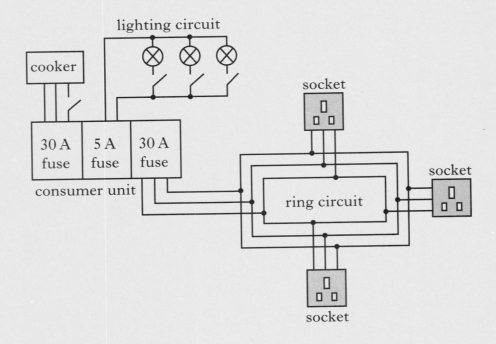

(a) (i) State **one** advantage of a ring circuit.

.... 1

(ii) State the value of mains voltage.

.... 1

(b) Each of the lamps in the lighting circuit has a power rating of 100 W. One of the lamps is switched on.

(i) Calculate the current in the lamp.

Space for working and answer

2

DO NOT WRITE IN THIS MARGIN

K&U	PS

Marks

5. (*b*) (continued)

(ii) Explain why a house with twenty 100 W lamps requires two separate lighting circuits.

....

.... **2**

[Turn over

6. A short-sighted person has difficulty seeing the picture on a cinema screen. Figure 1 shows rays of light from the screen entering an eye of the person until the rays reach the retina.

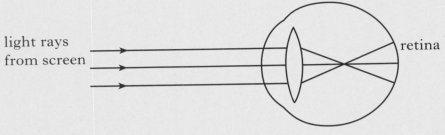

Figure 1

(a) (i) In the dotted box in Figure 2, draw the shape of lens that would correct this eye defect.

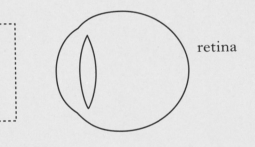

Figure 2

(ii) In Figure 2, complete the path of the rays of light from this lens until they reach the retina.

6. (continued)

(b) Doctors can use an endoscope to examine internal organs of a patient. The endoscope has two separate bundles of optical fibres that are flexible.

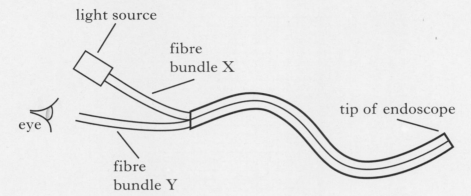

A section of optical fibre used in the endoscope is shown below.

(i) Complete the diagram to show how light is transmitted along the optical fibre. **2**

(ii) Explain the purpose of each bundle of optical fibres in the endoscope.

Fibre bundle X ....

....

Fibre bundle Y ....

.... **2**

(iii) The tip of the endoscope that is inside the patient is designed to be very flexible. Suggest **one** reason for this.

.... **1**

[Turn over

7. A hospital technician is working with a radioactive source. The graph shows the activity of the source over a period of time.

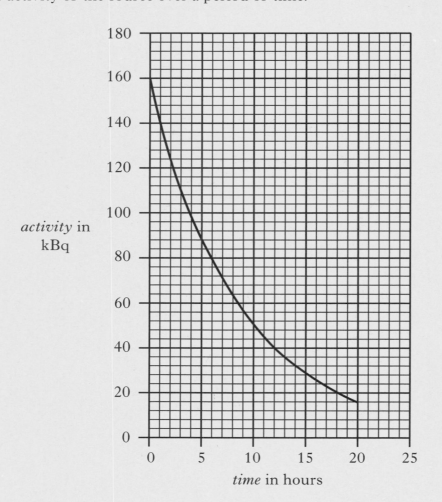

(a) (i) State what is meant by the term *half-life*.

....

(ii) Use information from the graph to calculate the half-life of the radioactive source.

Space for working and answer

7. **(a)** **(continued)**

(iii) The initial activity of the source is 160 kBq.

Calculate the activity, in kBq, of the radioactive source after four half-lives.

Space for working and answer

(b) As a safety precaution the technician wears a film badge when working with radioactive sources. The film badge contains photographic film. Light cannot enter the badge.

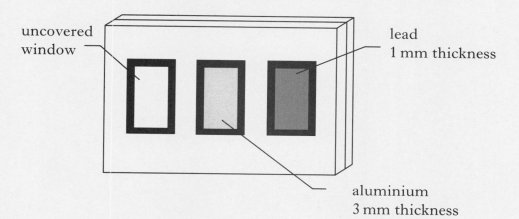

Describe how the film badge indicates the **type** and **amount** of radiation received.

....

....

....

....

[Turn over

8. A torch contains five identical LEDs connected to a 3·0 V battery as shown.

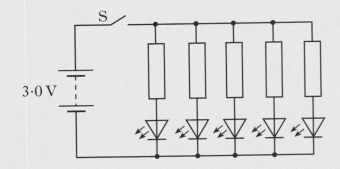

(a) State the purpose of the resistor connected in series with each LED.

....

(b) When lit, each LED operates at a voltage of 1·8 V and a current of 30 mA.

(i) Calculate the value of the resistor in series with each LED.

Space for working and answer

(ii) Calculate the total current from the supply when all five LEDs are lit.

Space for working and answer

8. **(b)** **(continued)**

(iii) Calculate the power supplied by the battery when all five LEDs are lit.

Space for working and answer

(c) State **one** advantage of using five LEDs rather than a single filament lamp in the torch.

....

[Turn over

9. An electronic device produces a changing light pattern when it detects music, but only when it is in the dark.

The device contains the logic circuit shown.

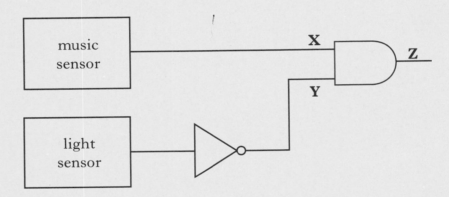

The music sensor produces logic 1 when the music is on and logic 0 when the music is off.

The light sensor produces logic 1 when it detects light and logic 0 when it is dark.

(a) (i) Suggest a suitable input device for the light sensor.

　... 1

(ii) Complete the truth table for the logic levels at points **X**, **Y** and **Z** in the circuit.

Music	Light level	X	Y	Z
off	dark			
off	light			
on	dark			
on	light			

3

9. (continued)

(b) The device detects music from a CD player. The CD player contains an amplifier that produces an output voltage of 5·6 V when connected to a loudspeaker of resistance 3·2 Ω.

   (i) Calculate the output power of the amplifier.

Space for working and answer

2

   (ii) The input power to the amplifier is 4·9 mW.
Calculate the power gain of the amplifier.

Space for working and answer

2

   (iii) One particular signal from the CD to the amplifier has a frequency of 170 Hz.
What is the frequency of the output signal from the amplifier?

....

1

[**Turn over**

10. A railway train travels uphill between two stations.

Information about the train and its journey is given below.

average speed of train	5 m/s
time for journey	150 s
power of train	120 kW
mass of train plus passengers	20 000 kg

(a) Calculate the energy used by the train during the journey.

Space for working and answer

10. (continued)

(b) Calculate the height gained by the train during the journey.

Space for working and answer

2

(c) Suggest why the actual height gained by the train is less than the value calculated in part (b).

....

.... **1**

[Turn over

11. A windsurfer takes part in a race. The windsurfer takes 120 seconds to complete the race. The total mass of the windsurfer and the board is 90 kg.

The graph shows how the speed of the windsurfer and board changes with time during part of the race.

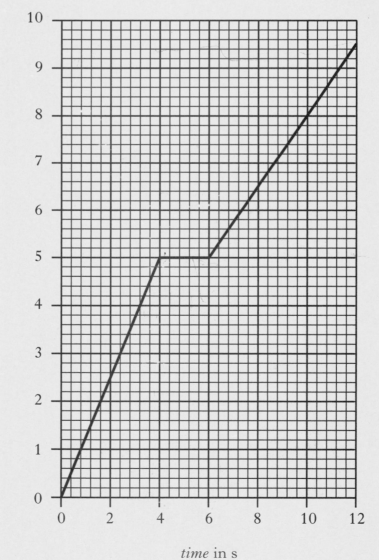

11. (continued)

(a) (i) Calculate the acceleration of the windsurfer and board during the first 4 s of the race.

Space for working and answer

2

(ii) Calculate the unbalanced force causing this acceleration.

Space for working and answer

2

(b) Calculate the total distance travelled by the windsurfer during the 12 s time interval shown on the graph.

Space for working and answer

2

(c) What can be said about the horizontal forces acting on the windsurfer between 4 s and 6 s?

....

1

[Turn over

12. An underwater generator is designed to produce electricity from water currents in the sea.

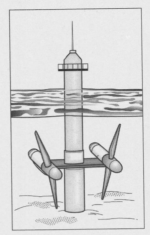

The output power of the generator depends on the speed of the water current as shown in Graph 1.

Graph 1

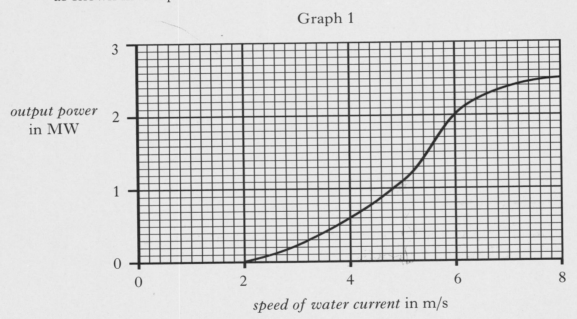

The speed of the water current is recorded at different times of the day shown in Graph 2.

Graph 2

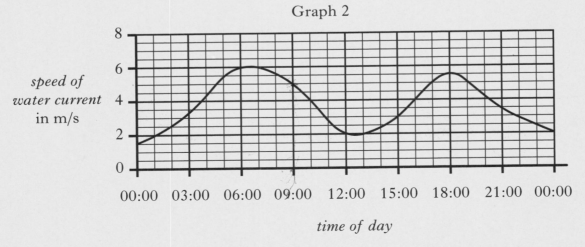

12. (continued)

Marks

(a) (i) State the output power of the generator at 09:00.

.... 1

(ii) State **one** disadvantage of using this type of generator.

.... 1

(b) The voltage produced by the generator is stepped-up by a transformer.

At one point in the day the electrical current in the primary coils of the transformer is 900 A and the voltage is 2000 V.

The transformer is 96% efficient.

(i) Calculate the output power of the transformer at this time.

Space for working and answer

3

(ii) State **one** reason why a transformer is not 100% efficient.

.... 1

[Turn over

12. (continued)

(c) Three different types of electrical generator, X, Y and Z are tested in a special tank with a current of water as shown to find out the efficiency of each generator.

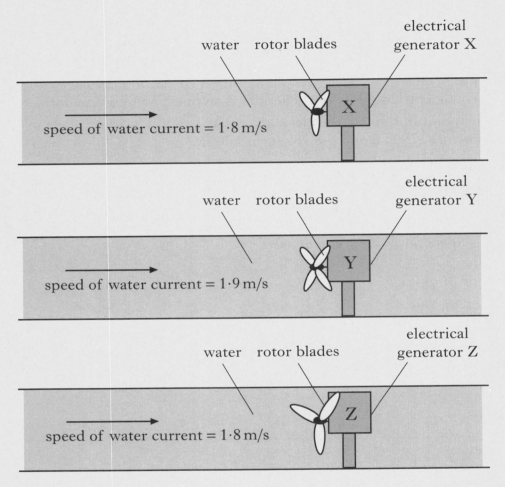

Give **two** reasons why this is not a fair test.

....

.... 1

13. In the reactor of a nuclear power station, neutrons split uranium nuclei to produce heat in what is known as a "chain reaction".

(a) Explain what is meant by the term "chain reaction".

....

.... 2

(b) In the nuclear power station, 1 kg of uranium fuel produces 4 200 000 MJ of heat. In a coal-fired power station 1 kg of coal produces 28 MJ of heat. Calculate how many kilograms of coal are required to produce the same amount of heat as 1 kg of uranium.

Space for working and answer

1

(c) A power station uses an a.c. generator to convert kinetic energy from a turbine into electrical energy. A diagram of an a.c. generator is shown.

(i) Explain how the a.c. generator works.

....

.... 2

(ii) State **two** changes that can be made to the generator to increase the output power.

Change 1: ....

Change 2: 2

14. A team of astronomers observes a star 200 light-years away.

(a) State what is meant by the term "light-year".

.... 1

(b) Images of the star are taken with three different types of telescope as shown.

| Telescope A | Telescope B | Telescope C |
| visible light | infrared | X-ray |

(i) Explain why different types of telescope are used to detect signals from space.

....

.... 2

(ii) Place the telescopes in order of the increasing wavelength of the radiation which they detect.

.... 1

(iii) State a detector that could be used in telescope C.

.... 1

(c) Telescope A is a refracting telescope with an objective lens of focal length 400 mm and diameter 80 mm.

(i) Calculate the power of the objective lens.

Space for working and answer

2

14. (c) (continued)

(ii) One of the astronomers suggests replacing the objective lens in this telescope with one of larger diameter.
State an advantage of doing this.

.... 1

[Turn over for Question 15 on *Page thirty*

15. (a) A spacecraft is used to transport astronauts and equipment to a space station. On its return from space the spacecraft must re-enter the Earth's atmosphere. The spacecraft has a heat shield made from special silica tiles to prevent the inside from becoming too hot.

(i) Why does the spacecraft increase in temperature when it re-enters the atmosphere?

....

(ii) The mass of the heat shield is 3.5×10^3 kg and the gain in heat energy of the silica tiles is 4.7 GJ.

Calculate the increase in temperature of the silica tiles.

Space for working and answer

(iii) Explain why the actual temperature rise of the silica tiles is less than the value calculated in (a)(ii).

....

....

(b) When a piece of equipment was loaded on to the spacecraft on Earth, two people were required to lift it.

One person was able to lift the same piece of equipment in the Space Station.

Explain why one person was able to lift the equipment in the Space Station.

....

[*END OF QUESTION PAPER*]

YOU MAY USE THE SPACE ON THIS PAGE TO REWRITE ANY ANSWER YOU HAVE DECIDED TO CHANGE IN THE MAIN PART OF THE ANSWER BOOKLET. TAKE CARE TO WRITE IN CAREFULLY THE APPROPRIATE QUESTION NUMBER.

[BLANK PAGE]

STANDARD GRADE | CREDIT

2009

OFFICIAL SQA PAST PAPERS   95   CREDIT PHYSICS 2009

FOR OFFICIAL USE

C

K&U   PS

3220/402

NATIONAL QUALIFICATIONS 2009

TUESDAY, 26 MAY 10.50 AM – 12.35 PM

PHYSICS STANDARD GRADE
Credit Level

Fill in these boxes and read what is printed below.

Full name of centre

Town

Forename(s)

Surname

Date of birth
Day Month Year   Scottish candidate number

Number of seat

Reference may be made to the Physics Data Booklet.

1 All questions should be answered.

2 The questions may be answered in any order but all answers must be written clearly and legibly in this book.

3 Write your answer where indicated by the question or in the space provided after the question.

4 If you change your mind about your answer you may score it out and rewrite it in the space provided at the end of the answer book.

5 If you use the additional space at the end of the answer book for answering any questions, you **must** write the correct question number beside each answer.

6 Before leaving the examination room you must give this book to the invigilator. If you do not, you may lose all the marks for this paper.

7 Any necessary data will be found in the **data sheet** on page three.

8 Care should be taken to give an appropriate number of significant figures in the final answers to questions.

Use **blue** or **black ink**. Pencil may be used for graphs and diagrams only.

SA 3220/402 6/18320

[BLANK PAGE]

DATA SHEET

Speed of light in materials

Material	Speed in m/s
Air	3.0×10^8
Carbon dioxide	3.0×10^8
Diamond	1.2×10^8
Glass	2.0×10^8
Glycerol	2.1×10^8
Water	2.3×10^8

Speed of sound in materials

Material	Speed in m/s
Aluminium	5200
Air	340
Bone	4100
Carbon dioxide	270
Glycerol	1900
Muscle	1600
Steel	5200
Tissue	1500
Water	1500

Gravitational field strengths

	Gravitational field strength on the surface in N/kg
Earth	10
Jupiter	26
Mars	4
Mercury	4
Moon	1.6
Neptune	12
Saturn	11
Sun	270
Venus	9

Specific heat capacity of materials

Material	Specific heat capacity in J/kg °C
Alcohol	2350
Aluminium	902
Copper	386
Glass	500
Glycerol	2400
Ice	2100
Lead	128
Silica	1033
Water	4180

Specific latent heat of fusion of materials

Material	Specific latent heat of fusion in J/kg
Alcohol	0.99×10^5
Aluminium	3.95×10^5
Carbon dioxide	1.80×10^5
Copper	2.05×10^5
Glycerol	1.81×10^5
Lead	0.25×10^5
Water	3.34×10^5

Melting and boiling points of materials

Material	Melting point in °C	Boiling point in °C
Alcohol	−98	65
Aluminium	660	2470
Copper	1077	2567
Glycerol	18	290
Lead	328	1737
Turpentine	−10	156

Specific latent heat of vaporisation of materials

Material	Specific latent heat of vaporisation in J/kg
Alcohol	11.2×10^5
Carbon dioxide	3.77×10^5
Glycerol	8.30×10^5
Turpentine	2.90×10^5
Water	22.6×10^5

SI Prefixes and Multiplication Factors

Prefix	Symbol	Factor	
giga	G	1 000 000 000	$= 10^9$
mega	M	1 000 000	$= 10^6$
kilo	k	1000	$= 10^3$
milli	m	0.001	$= 10^{-3}$
micro	μ	0.000 001	$= 10^{-6}$
nano	n	0.000 000 001	$= 10^{-9}$

1. A laptop computer uses a radio signal to transfer information to a base station. The base station is connected by optical fibres to a telephone exchange.

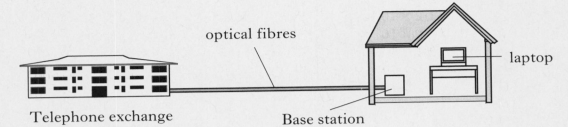

(a) The frequency of the radio signal is 5 GHz.

(i) State the speed of the radio signal.

....

(ii) Calculate the wavelength of the radio signal.

Space for working and answer

(b) The telephone exchange is 40 km away from the base station.

Calculate the time taken for the signal to travel along the **glass** optical fibre from the base station to the local telephone exchange.

Space for working and answer

(c) Copper wire can also be used to transfer information between the base station and the telephone exchange.

State **one** advantage of using optical fibres compared to copper wire.

....

2. A ship is carrying out a survey of the sea bed using ultrasound waves.

When stationary, the ship transmits and receives pulses of ultrasound waves. The transmitted ultrasound waves have a frequency of 30 kHz.

(a) What is meant by ultrasound?

.... **1**

(b) What is the speed of ultrasound waves in water?

.... **1**

(c) One pulse of ultrasound is received back at the ship 0·36 s after being transmitted.

Calculate the depth of the sea bed.

Space for working and answer

3

[Turn over

3. A rock concert is being held at Hampden Stadium. The concert is being filmed and is displayed on a large screen above the stage. This allows the band to be seen clearly by people at the back of the stadium.

(a) The people at the back of the stadium, watching the screen, notice that there is a time delay between seeing the drummer hitting the drums and hearing the sound.

Explain why there is a time delay.

....

.... 1

(b) The concert is also being broadcast live on radio and television.

The audio signal is combined with a radio carrier wave to produce a modulated radio signal.

The audio signal and the modulated radio signal are shown below.

Draw the radio carrier wave in the space provided.

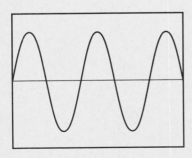
audio signal

radio carrier wave

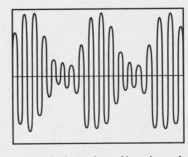
modulated radio signal

2

3. (continued)

(c) An electric guitar used in the concert is connected to an amplifier.

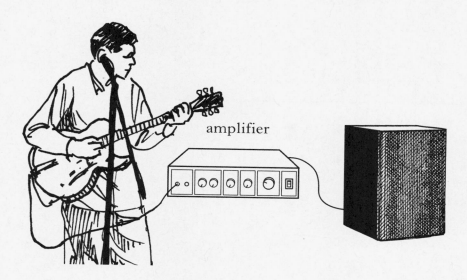

The input power of the signal from the guitar to the amplifier is 30 mW.
The output of the amplifier is connected to a loudspeaker.
The amplifier has a power gain of 25 000.

Calculate the output power delivered to the loudspeaker.

Space for working and answer

2

[Turn over

4. A car fan uses a battery powered electric motor. The diagram below shows the apparatus used to investigate the effect of current on the speed of the electric motor.

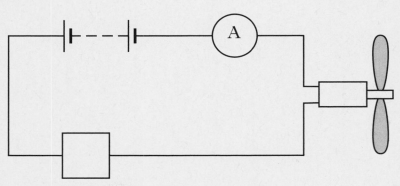

variable speed control

(a) The graph shows the relationship between speed and current during the investigation.

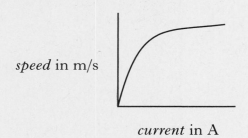

(i) The current is changed using the variable speed control.

What happens to the current when the resistance of the variable speed controller is reduced?

....

(ii) The settings of the variable speed control use different combinations of **identical** resistors, as shown.

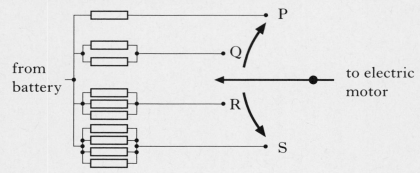

(A) To which position should the variable speed control be set to achieve maximum speed?

....

(B) Justify your answer.

....

4. (continued)

(b) The electric motor is shown below.

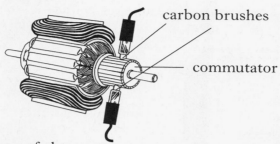

(i) Explain the purpose of the commutator.

....

.... 1

(ii) Why are the brushes made of carbon rather than metal wire?

.... 1

(c) When a wire carrying a current is placed in a magnetic field, a force is produced on the wire. The diagram shows the direction of the force for a particular situation.

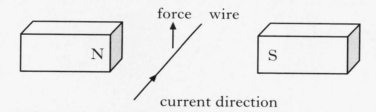

(i) A simplified diagram of an electric motor is shown below. Indicate on the diagram the direction of the force on the wire at point X and point Y. 1

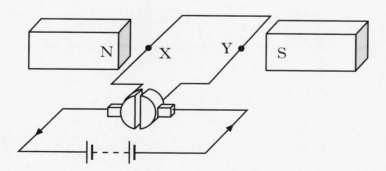

(ii) State **one** way in which the direction of rotation of the motor could be reversed.

.... 1

5. A householder plugs a home entertainment centre, a hi-fi, a games console and an electric fire into a multiway adaptor connected to the mains.

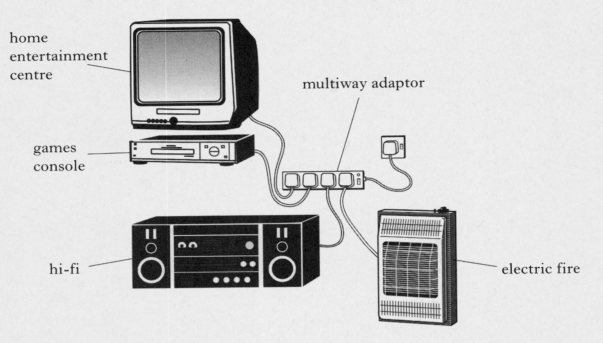

The wiring in the electric fire is found to be faulty. The circuit is shown below.

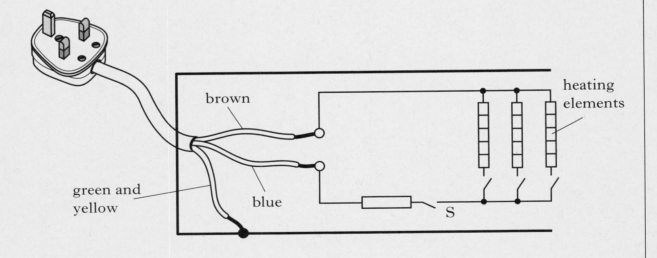

(a) What is the fault in the circuit?

....

.... 1

5. (continued)

(b) The householder goes on holiday for 14 days.

The electric fire is unplugged.

All the other appliances are left on standby.

On standby, these appliances operate at 9·0% of their power rating listed in the table.

Appliance	Power rating (W)
home entertainment centre	350
hi-fi	150
games console	300
electric fire	2080

(i) Calculate the total power consumption, in watts, of all the appliances left **on standby**.

Space for working and answer

(ii) Calculate the number of kilowatt-hours used by these appliances during the 14 days on standby.

Space for working and answer

[Turn over

6. In a physics laboratory, a student wants to find the focal length of a convex lens. The student is given a sheet of white paper, a metre stick and a lens.

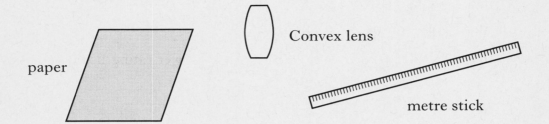

(a) Explain how the student could measure the focal length of the lens using this equipment.

....

.... **2**

(b) Refraction of light occurs in lenses.

What is meant by the term refraction?

....

.... **2**

(c) The following diagram shows a ray of light entering a glass block.

(i) Complete the diagram to show the path of the ray of light through the block and after it emerges from the block.

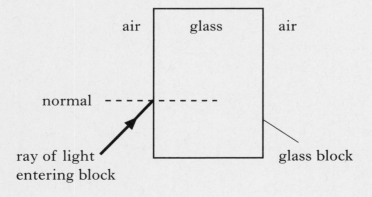

2

(ii) On your diagram indicate an angle of refraction, **r**. **1**

7. Students observe an experiment with radioactive sources. The radiation is measured using a detector and counter. The background count rate is measured.

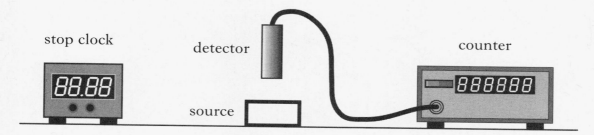

Different absorbing materials are then placed, in turn, between source and detector and readings for each material are recorded. This is repeated for each source. The results are shown in the table.

Source	Corrected Count Rate (Counts per minute)			
	No absorbing material	Paper	2 cm of Aluminium	2 cm of Lead
A	480	480	480	200
B	720	300	300	180
C	600	580	0	0

One source emits beta radiation only, one emits gamma only and one emits both alpha and gamma radiation.

(a) Complete the following table to identify the source.

Type of radiation	Source
beta only	
both alpha and gamma	

(b) One source has a half-life of 30 minutes.
The source has an initial activity of 18 000 Bq.
Calculate its activity after 2 hours.

8. A digital camera is used to take pictures. When switched on, the flash on a digital camera requires some time before it is ready to operate. When ready, a green LED is illuminated.

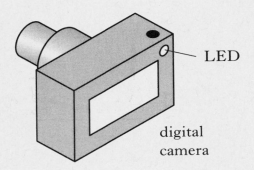

digital camera

The part of the circuit used to control the LED is shown below. The voltage at point X is initially 0 V.

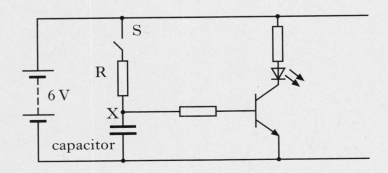

(a) Describe what happens to the voltage at point X when switch S is closed.

.... 1

(b) The camera manufacturer wants to change the time taken for the flash to be ready to operate.

State **two** changes which could be made to the above circuit so that the time for the green LED to come on is **reduced**.

....

.... 2

8. (continued)

(c) The camera flash is designed to operate under dim lighting conditions. Another part of the circuit for the camera flash is shown below. The flash only operates when a minimum voltage of 0·7 V occurs across the LDR.

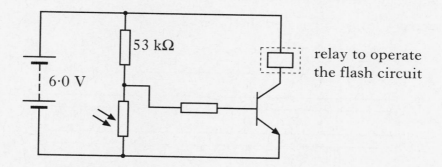

(i) Calculate the voltage across the 53 kΩ resistor when the voltage across the LDR is 0·7 V.

Space for working and answer

1

(ii) Calculate the **minimum** resistance of the LDR that allows the flash to operate in dim conditions.

Space for working and answer

2

[Turn over

9. A remote gas sensing system detects and identifies whether hydrogen, helium and oxygen gases are present in a sample.

Sensors, consisting of light detectors with filters in front of them, are linked to a processing system that can provide a recognisable output to identify each gas.

The filters allow a limited band of wavelengths to pass through them.

The line spectrum for each gas and the position of filters A, B, C and D are shown below.

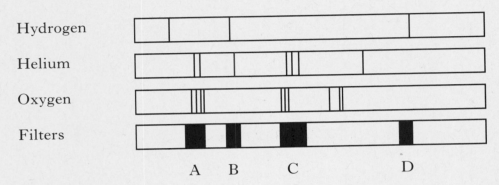

If a spectral line is at the same position as a filter band then that sensor will produce a logic level one.

(a) Suggest a suitable input device for the sensor.

....

(b) Complete the truth table for the sensor outputs when each gas is detected. Hydrogen has already been completed.

Gas	Sensor			
	A	B	C	D
Hydrogen	0	1	0	1
Helium				
Oxygen				

9. (continued)

(c) The logic circuit used to identify one of these gases is shown.

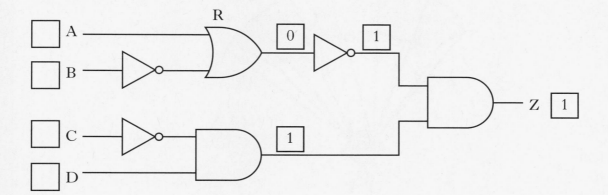

(i) Name logic gate R.

.... 1

(ii) When this gas is present, a logic 1 is output at Z.

(A) Complete the boxes A, B, C and D in the logic circuit.

(B) Name which gas is detected with this circuit.

.... 2

[Turn over

10. A parachutist jumps out of an aircraft. Sometime later, the parachute is opened.

The graph shows the motion of the parachutist from leaving the aircraft until landing.

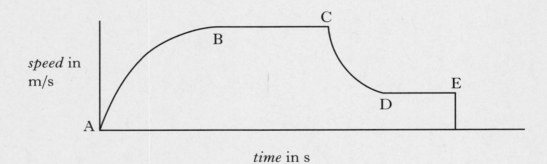

(a) Which parts of the graph show when the forces acting on the parachutist are balanced?

.... 1

10. (continued)

(b) The parachutist lands badly and is airlifted to hospital by helicopter.

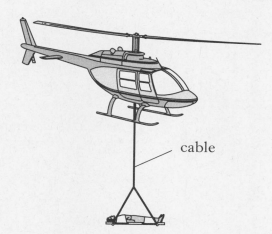

The stretcher and parachutist have a total mass of 90·0 kg.

(i) Calculate the weight of the stretcher and parachutist.

Space for working and answer

(ii) The helicopter cable provides an upward force of 958·5 N to lift the stretcher and parachutist.

Calculate the acceleration of the stretcher and parachutist.

Space for working and answer

11. Two students set up a linear air track experiment. A linear air track consists of a hollow tube with small holes. Air is blown through the small holes. A vehicle moves on the cushion of air.

   The vehicle starts from rest at **X** and moves along the air track so that the card passes through the light gate at point **Y**.

   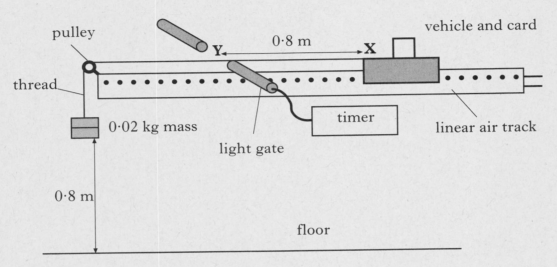

   The results for one experiment are recorded in the table below.

Card Length (cm)	Speed at **X** (m/s)	Timer Reading at **Y** (s)	Speed at **Y** (m/s)	Time to travel from **X** to **Y** (s)	Acceleration between **X** and **Y** (m/s^2)
3	0	0·05	0·6	1·5	

   (a) Use the information given in the table to calculate the acceleration of the vehicle between **X** and **Y**.

   Space for working and answer

   2

DO NOT WRITE IN THIS MARGIN

K&U | PS

Marks

11. (continued)

(*b*) When repeating the experiment, the 0·02 kg mass detaches from the thread before the vehicle is released. The mass falls 0·80 m to the floor.

(i) Calculate the gravitational potential energy stored in the mass before it fell.

Space for working and answer

2

(ii) Assuming the mass falls from rest, calculate the final speed of the mass just before it hits the floor.

Space for working and answer

3

[Turn over

12. A hovercraft service was trialled on the Firth of Forth from Kirkcaldy to Leith.

The hovercraft and passengers have a total weight of 220 000 N.

(a) State the value of the upward force exerted on the hovercraft when it hovers at a constant height.

....

(b) The graph shows how the speed of the hovercraft varies with time for one journey from Kirkcaldy to Leith.

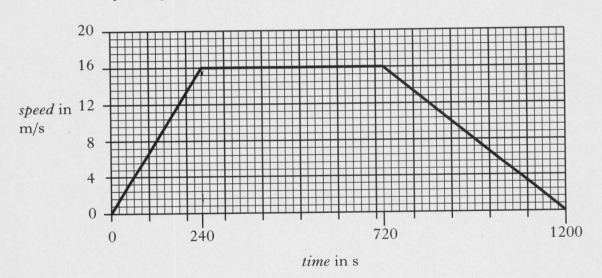

(i) Calculate the total distance travelled during the journey.

Space for working and answer

DO NOT WRITE IN THIS MARGIN

K&U	PS

Marks

12. (*b*) (continued)

(ii) Calculate the average speed for the whole journey.

Space for working and answer

2

[Turn over

13. The National Grid transfers electrical energy across the country from power stations using a 132 kV network. Electrical power is generated at 20 kV and 5 kA from the power station generator, before being increased to 132 kV using a transformer.

(a) What is the reason for increasing the voltage of the electrical power?

....

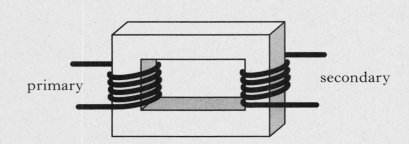

(b) There are 2000 turns in the primary circuit of the transformer. Assuming the transformer is 100% efficient:

(i) calculate the number of turns in the secondary coil;

Space for working and answer

(ii) calculate the current in the secondary coil of the transformer.

Space for working and answer

13. (continued)

(c) The secondary coil of the transformer is connected to the high voltage National Grid network. High voltage cable has a resistance of 0·31 Ω/km. One cable has a length of 220 km.

Calculate the power loss in this cable.

Space for working and answer

3

[Turn over

14. A solar furnace consists of an array of mirrors which reflect heat radiation on to a central curved reflector.

A heating container is placed at the focus of the central curved reflector. Metals placed in the container are heated until they melt.

The diagram below shows the heat rays after reflection by the mirrors on the hillside.

(a) Complete the diagram to show the effect of the central curved reflector on the heat rays.

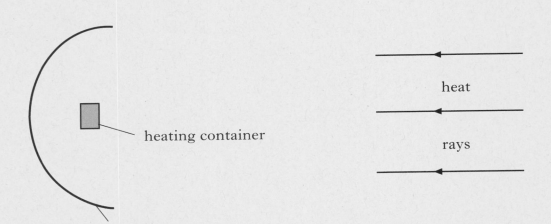

14. (continued)

(b) 8000 kg of pre-heated aluminium pellets at a temperature of 160 °C are placed in the container. Aluminium has a specific heat capacity of 902 J/kg °C and a melting point of 660 °C.

How much heat energy is required to heat the aluminium to its melting point?

Space for working and answer

2

(c) (i) How much extra energy is required to melt the aluminium pellets?

Space for working and answer

3

(ii) The power of the furnace is 800 kW. How long will it take for this extra energy to be supplied?

Space for working and answer

2

(iii) Explain why it takes longer, in practice, to melt the aluminium.

....

....

1

[Turn over

15. A space probe is designed to record data on its way to landing on Ganymede, a moon of Jupiter. The launch vehicle is made up of the probe of mass 8000 kg and the constant thrust rocket unit which has a mass of 117 000 kg.

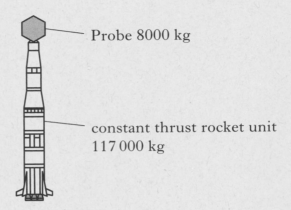

- Probe 8000 kg
- constant thrust rocket unit 117 000 kg

On launch, the resultant force acting upwards on the launch vehicle is 1 400 000 N.

(a) Calculate the initial acceleration of the launch vehicle.

Space for working and answer

(b) As the launch vehicle continues to ascend, its acceleration increases. This is partly due to the decrease in gravitational field strength as it gets further from Earth.

Give another reason why the acceleration increases.

....

15. (continued)

(c) The space probe eventually goes into orbit around Ganymede.

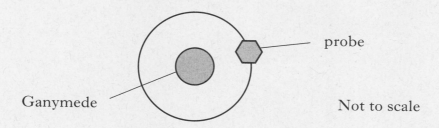

Not to scale

Explain why the probe follows a circular path while in orbit.

....

....

.... **2**

(d) The probe has gas thrusters that fire to slow it down in order to land on Ganymede. In terms of Newton's laws, explain how these thrusters achieve this task.

....

....

.... **2**

[Turn over

16. (*a*) Astronomers use refracting telescopes to observe planets. A refracting telescope has an eyepiece lens and an objective lens.

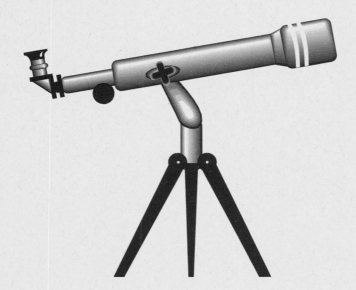

(i) An eyepiece lens can be used on its own as a magnifying glass. Complete the ray diagram to show how the eyepiece lens forms a magnified image.

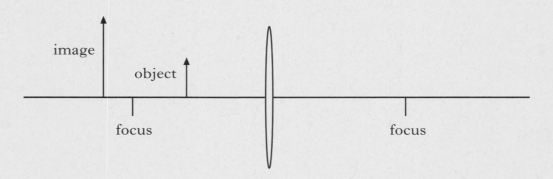

(ii) How does the diameter of the objective lens affect the image seen through the telescope?

....

16. (continued)

(b) Radio waves emitted by galaxies are detected and used to provide images of the galaxies.

(i) How does the wavelength of radio waves compare with the wavelength of light?

.... 1

(ii) Name a detector for radio waves.

.... 1

(iii) Why are different kinds of telescope used to detect signals from space?

....

....

.... 1

[END OF QUESTION PAPER]

ADDITIONAL SPACE FOR ANSWERS

Make sure you write the correct question number beside each answer.

STANDARD GRADE | CREDIT

2010

OFFICIAL SQA PAST PAPERS   129   CREDIT PHYSICS 2010

FOR OFFICIAL USE

K&U   PS

3220/402

NATIONAL QUALIFICATIONS 2010

FRIDAY, 28 MAY 10.50 AM – 12.35 PM

PHYSICS STANDARD GRADE Credit Level

Fill in these boxes and read what is printed below.

Full name of centre

Town

Forename(s)

Surname

Date of birth
Day Month Year

Scottish candidate number

Number of seat

Reference may be made to the Physics Data Booklet.

1 All questions should be answered.

2 The questions may be answered in any order but all answers must be written clearly and legibly in this book.

3 Write your answer where indicated by the question or in the space provided after the question.

4 If you change your mind about your answer you may score it out and rewrite it in the space provided at the end of the answer book.

5 If you use the additional space at the end of the answer book for answering any questions, you **must** write the correct question number beside each answer.

6 Before leaving the examination room you must give this book to the Invigilator. If you do not, you may lose all the marks for this paper.

7 Any necessary data will be found in the **data sheet** on page three.

8 Care should be taken to give an appropriate number of significant figures in the final answers to questions.

Use **blue** or **black ink**. Pencil may be used for graphs and diagrams only.

[BLANK PAGE]

DATA SHEET

Speed of light in materials

Material	Speed in m/s
Air	3.0×10^8
Carbon dioxide	3.0×10^8
Diamond	1.2×10^8
Glass	2.0×10^8
Glycerol	2.1×10^8
Water	2.3×10^8

Speed of sound in materials

Material	Speed in m/s
Aluminium	5200
Air	340
Bone	4100
Carbon dioxide	270
Glycerol	1900
Muscle	1600
Steel	5200
Tissue	1500
Water	1500

Gravitational field strengths

	Gravitational field strength on the surface in N/kg
Earth	10
Jupiter	26
Mars	4
Mercury	4
Moon	1.6
Neptune	12
Saturn	11
Sun	270
Venus	9

Specific heat capacity of materials

Material	Specific heat capacity in J/kg °C
Alcohol	2350
Aluminium	902
Copper	386
Glass	500
Glycerol	2400
Ice	2100
Lead	128
Silica	1033
Water	4180

Specific latent heat of fusion of materials

Material	Specific latent heat of fusion in J/kg
Alcohol	0.99×10^5
Aluminium	3.95×10^5
Carbon dioxide	1.80×10^5
Copper	2.05×10^5
Glycerol	1.81×10^5
Lead	0.25×10^5
Water	3.34×10^5

Melting and boiling points of materials

Material	Melting point in °C	Boiling point in °C
Alcohol	−98	65
Aluminium	660	2470
Copper	1077	2567
Glycerol	18	290
Lead	328	1737
Turpentine	−10	156

Specific latent heat of vaporisation of materials

Material	Specific latent heat of vaporisation in J/kg
Alcohol	11.2×10^5
Carbon dioxide	3.77×10^5
Glycerol	8.30×10^5
Turpentine	2.90×10^5
Water	22.6×10^5

SI Prefixes and Multiplication Factors

Prefix	Symbol	Factor	
giga	G	1 000 000 000	$= 10^9$
mega	M	1 000 000	$= 10^6$
kilo	k	1000	$= 10^3$
milli	m	0.001	$= 10^{-3}$
micro	μ	0.000 001	$= 10^{-6}$
nano	n	0.000 000 001	$= 10^{-9}$

1. A car is fitted with a parking system. This warns the driver how close objects are behind the car. Equipment on the rear bumper of the car transmits ultrasound waves and receives the reflected waves.

(a) (i) Use the data sheet to find the speed of ultrasound waves in air.

....

(ii) The ultrasound waves have a frequency of 40 kHz.

Calculate the wavelength of these waves.

Space for working and answer

(b) The car stops 1·7 m from a wall.

Calculate the time for a transmitted wave to return to the car.

Space for working and answer

(c) The car is moved closer to the wall.

State what happens to the time for a transmitted wave to return to the car.

....

2. A hill lies between a radio and television transmitter and a house.

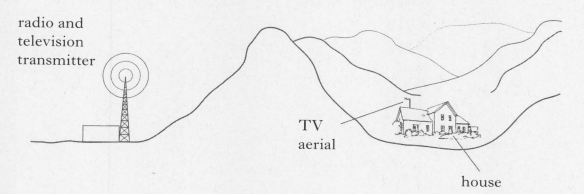

The house is within the range of both the radio and television signals from the transmitter.

(a) In the house, a radio has good reception but a TV has poor reception from this transmitter.

Suggest an explanation for this.

....

....

....

....

(b) The house is also fitted with a dish aerial to receive TV signals from a geostationary satellite. The TV signals are carried by microwaves with a frequency of 12 GHz.

(i) State the speed of microwave signals in air.

....

(ii) What is meant by the term geostationary?

....

....

[Turn over

3. The circuit shown is used to control the brightness of two **identical** lamps. The variable resistor is adjusted until the lamps operate at their correct voltage of 3·0 V.

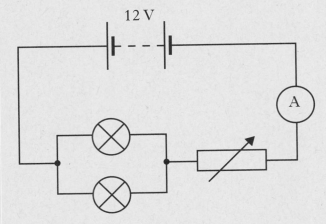

(a) When the lamps operate at the correct voltage, the reading on the ammeter is 1·2 A.

Calculate the current in one lamp.

Space for working and answer

(b) Calculate the resistance of one lamp.

Space for working and answer

3. (continued)

(c) Calculate the combined resistance of the two lamps in this circuit.

Space for working and answer

(d) When the lamps operate at their correct voltage the resistance of the variable resistor is $7.5\,\Omega$.

Calculate the total resistance in the circuit.

Space for working and answer

(e) One of the lamps is removed.

(i) What happens to the reading on the ammeter?

....

(ii) Justify your answer.

....

[Turn over

DO NOT WRITE IN THIS MARGIN

K&U PS

Marks

4. A washing machine contains a commercial electric motor. The rating plate on the washing machine shows the following information.

2530 W
230 V ac
50 Hz

The plug connected to the washing machine contains a 13 A fuse.

(*a*) (i) State the purpose of the fuse.

.... **1**

(ii) Show by calculation that a 3 A fuse is unsuitable.

Space for working and answer

2

4. (continued)

(b) A student builds a simple dc electric motor. Some differences between a commercial motor and a simple dc motor are shown in the table.

Commercial motor	Simple dc motor
field coils	permanent magnets
multi-section commutator	commutator
carbon brushes	brushes

State a reason for a commercial motor using:

(i) field coils instead of permanent magnets;

.... 1

(ii) a multi-section commutator instead of a single commutator.

.... 1

(c) The electrical energy used by a commercial motor is measured in kilowatt-hours.

Calculate how many joules are equivalent to one kilowatt-hour.

Space for working and answer

2

[Turn over

5. In the eye, refraction of light occurs at both the cornea and the lens.

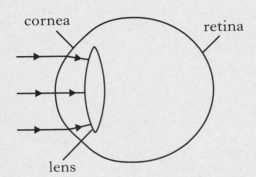

(a) The focal length of an eye lens system (the cornea and the lens together) is 22 mm.

Calculate the power of this eye lens system.

Space for working and answer

5. (continued)

(b) A student has an eye defect. An object close to the student's eye appears focused but a distant object appears blurred.

(i) What name is given to this eye defect?

.... 1

(ii) The diagram shows rays of light, from a distant object, entering the student's eye.

Complete the diagram to show how the light rays reach the retina of the student's eye. 1

(iii) By referring to your completed diagram, explain why the image on the retina of the student's eye is blurred.

.... 1

(iv) A lens is used to correct this eye defect.

Draw the shape of this lens.

Space for drawing

1

[Turn over

6. The table gives information about radioactive substances used in medicine.

Radioactive substance	Type of ionising radiation emitted	Half-life
iodine-131	beta and gamma	8 days
technetium-99 m	gamma	6 hours
cobalt-60	beta and gamma	5·3 years

(a) (i) State what is meant by the term ionisation.

....

....

(ii) State a type of ionising radiation **not** given in the table above.

....

(b) A sample of iodine-131 is delivered to a hospital 24 days before it is given to a patient. The activity of the iodine-131 when it is given to the patient is 6 MBq.

Calculate the initial activity, in MBq, of the sample when it was delivered to the hospital.

Space for working and answer

6. (continued)

   (c) (i) Equivalent dose measures the biological effect of radiation.

   State the unit of equivalent dose.

   1

   (ii) For living material the biological effect of radiation depends on a number of factors.

   State **two** of these factors.

   1 ....

   2 2

   [Turn over

7. A physics student builds a lap counter for a toy racing car set. The lap is counted when the car passes over the light sensor.

(a) The circuit for the light sensor contains an LDR as shown.

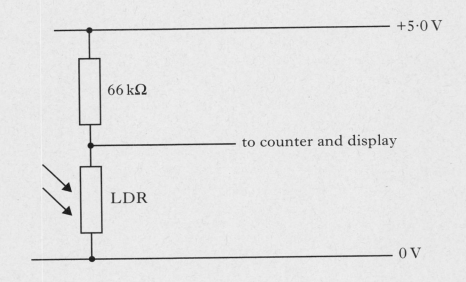

The resistance of the LDR for different conditions is shown in the table.

Light Sensor	Resistance of LDR (kΩ)
covered	22
uncovered	2

7. **(a)** **(continued)**

Calculate the voltage across the LDR when the light sensor is covered.

Space for working and answer

3

(b) (i) The system contains a counter and display. The output of the counter is **binary**. This is then converted to **decimal** and shown on the display. What decimal number is shown when the counter output is 1001?

....

1

(ii) The system also contains a buzzer. The buzzer emits a sound when a car completes a lap. The buzzer has a resistance of 120 Ω and a power of 147 mW.

Calculate the voltage across the buzzer when it sounds.

Space for working and answer

2

[Turn over

8. An electronic device warns a car driver when the seat belt has not been fastened. The device only operates when the ignition is switched on. The device contains the logic circuit shown.

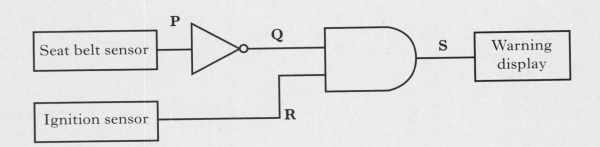

The seat belt sensor produces logic 1 when the seat belt is fastened and logic 0 when the seat belt is unfastened.

The ignition sensor produces logic 1 when the car ignition is on and logic 0 when the car ignition is off.

(a) (i) Suggest a suitable output device that will illuminate the warning display.

....

(ii) Complete the truth table for the logic levels **P**, **Q** and **S** in the circuit.

Seat belt	Ignition	P	Q	R	S
unfastened	off			0	
unfastened	on			1	
fastened	off			0	
fastened	on			1	

(b) Explain in terms of forces, why seat belts are used in cars.

....

....

....

DO NOT WRITE IN THIS MARGIN

K&U PS

Marks

8. (continued)

(*c*) The car has another electronic device that also contains a logic gate. The truth table for **this** logic gate is shown below.

Input 1	Input 2	Output
0	0	0
0	1	1
1	0	1
1	1	1

(i) Name this logic gate.

.... **1**

(ii) Draw the symbol for this logic gate.

Space for symbol

1

(*d*) The temperature outside the car is measured with an electronic thermometer and displayed on a screen.

What input device could be used in the electronic thermometer?

.... **1**

[Turn over

9. A student releases a trolley from rest near the top of a track. The trolley moves down the track. A card attached to the trolley passes through a light gate near the bottom of the track.

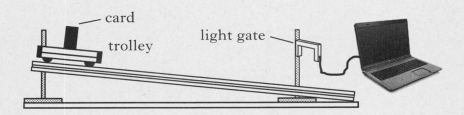

(a) The student records the following information.

Length of the card = 60 mm
Distance travelled by the trolley down the track = 1·2 m
Time for the card to pass through the light gate = 0·075 s

Calculate the instantaneous speed of the trolley as it passes through the light gate.

Space for working and answer

(b) The mass of the trolley is 0·55 kg.

Calculate its kinetic energy as it passes through the light gate.

Space for working and answer

(c) Suggest a possible value for the average speed of the trolley over the 1·2 m distance travelled by the trolley down the track.

....

10. A cyclist is approaching traffic lights at a constant speed. The cyclist sees the lights change to red. The graph shows how the speed of the cyclist varies with time from the instant the cyclist sees the lights change to red.

(a) (i) How long did it take for the cyclist to react before applying the brakes?

....

(ii) Calculate the distance travelled from the instant the cyclist sees the traffic lights change to red until stationary.

Space for working and answer

(b) The cyclist now sees the traffic lights change to green and accelerates away from the lights. The combined mass of the cycle and cyclist is 75 kg. An unbalanced force of 150 N acts on this combined mass.

Calculate the acceleration.

Space for working and answer

11. A wind generator is used to charge a 12 V battery. The charging current depends on the wind speed.

The graph shows the charging current at different wind speeds.

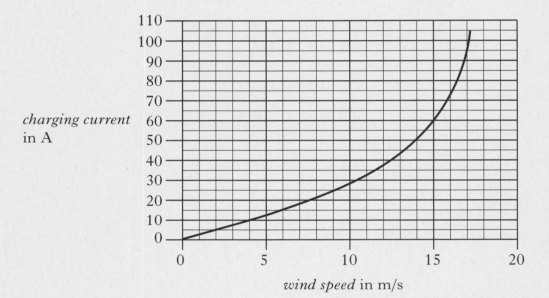

wind speed in m/s

(a) During one charge of the battery, the wind speed is constant at 15 m/s. During this time a charge of 4500 C is transferred to the battery.

Calculate the time taken to transfer this charge to the battery.

Space for working and answer

3

11. (continued)

(b) At another wind speed the generator has an output power of 120 W and is 30% efficient.

Calculate the input power to the generator.

Space for working and answer

2

(c) A bicycle has a small generator called a dynamo. The dynamo contains a magnet which spins near a coil of wire.

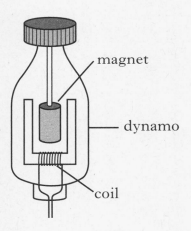

When the magnet spins, a voltage is induced in the coil.

State **two** factors that affect the size of the induced voltage.

1 ....

2 ....

2

[Turn over

12. A technician tests an electric kettle. The kettle is filled with water and switched on for 3 minutes.

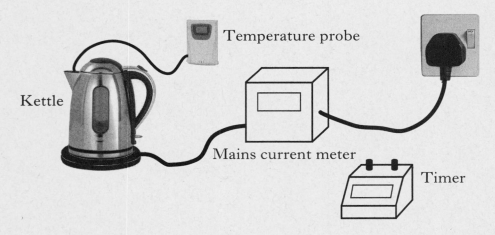

The technician records the following information.

| Current = 12·5 A |
| Voltage = 230 V |
| Time = 3 minutes |
| Initial temperature = 18 °C |
| Final temperature = 90 °C |

(a) (i) Show that 517 500 J of electrical energy is supplied to the kettle in 3 minutes.

Space for working and answer

2

12. (*a*) (**continued**)

    (ii) Calculate the mass of water in the kettle.

Space for working and answer

    (iii) Explain why the mass of water will be less than calculated in (*a*)(ii).

....

(*b*) The technician tests a second kettle. When the water boils this kettle does not switch off and continues to heat the water.

    (i) State what happens to the temperature of the water when it boils.

....

While the water is boiling, the kettle supplies 565 000 J of heat energy to the water.

    (ii) Calculate the mass of water changed into steam.

Space for working and answer

[Turn over

13. During a visit to a science centre a student learns more about gravitational field strength.

(a) State what is meant by gravitational field strength.

....

....

(b) The science centre has a set of specially designed scales. The weight of the student on different planets in the solar system can be found by using these scales. The student stands on each of these scales in turn. The weight on each of these scales is shown.

Mercury 280 N   Planet X 630 N   Neptune 840 N

(i) The student has a mass of 70 kg.

Calculate the gravitational field strength on Planet X.

Space for working and answer

(ii) Identify Planet X.

....

13. (continued)

(c) The student watches a short film. This film shows an astronaut dropping a hammer onto the surface of the Moon. The hammer takes 1·2 s to fall to the Moon's surface. The gravitational field strength on the Moon is 1·6 N/kg.

(i) Calculate the vertical speed of the hammer just before it strikes the Moon's surface.

Space for working and answer

2

(ii) The film then shows the astronaut throwing the hammer horizontally from the same height.

How long does it take for the hammer to fall to the Moon's surface?

.... 1

[Turn over

14. The diagram represents the electromagnetic spectrum.

Some of the radiations have not been named.

| gamma | X-rays | P | Visible light | Q | microwaves | Radio and TV |

Electromagnetic spectrum

← increasing frequency

(a) (i) Name radiations P and Q.

P ....

Q   2

(ii) Which radiation in the electromagnetic spectrum has the shortest wavelength?

....   1

(iii) State **one** detector of radio waves.

....   1

(iv) State **one** medical use of infrared radiation.

....   1

(b) Yellow light is part of the visible spectrum. The wavelength of yellow light is 5.9×10^{-7} m.

The visible spectrum also contains red, blue and green light.

Use the information above to complete the following table.

Colour	Wavelength (m)
	7×10^{-7}
yellow	
	5.5×10^{-7}
	4.5×10^{-7}

2

14. (continued)

(c) The table below gives information about planets that orbit the Sun.

Planet	Distance from the Sun (Gm)	Period (days)	Mass (Earth masses)
Earth	150	365	1
Jupiter	780		318
Mars	228	687	0·11
Mercury	58	88	0·06
Saturn	1430	10 760	95
Venus	110	225	0·82

(i) Give an approximate value, **in days**, for the period of Jupiter.

....

(ii) Calculate the time taken for light from the Sun to reach Saturn.

Space for working and answer

[END OF QUESTION PAPER]

ADDITIONAL SPACE FOR ANSWERS

Make sure you write the correct question number beside each answer.